Sustainable Development

Promoting sustainable development opens up debates about our relationship with the natural world, about what constitutes social progress and about the character of development, both in the North and the South, in the present and into the future.

This concise and accessible text explores the prospects for and barriers to the promotion of sustainable development in the high-consumption societies of the industrialized world, the Third World and the economies in transition in East and Central Europe. Sustainable development is explored as part of new efforts, albeit tentative, to integrate environmental, economic and (more recently) social considerations into a new development paradigm. Recognizing that promoting sustainable development is a quintessentially global task, this valuable book focuses on the authoritative Brundtland formulation of sustainable development and the role of the United Nations Summits in promoting this vision.

Drawing on a myriad of global case studies such as Central Africa, India and New Zealand, this engaging book introduces students to the issues involved in the promotion of sustainable development in a theoretically informed and critical way.

Susan Baker is Professor in Environmental Social Sciences, Cardiff School of Social Sciences, Cardiff University, Wales.

Routledge introductions to environment series
Published and forthcoming titles

Titles under Series Editors:
Rita Gardner and A.M. Mannion

Environmental Science texts

Atmospheric Processes and Systems
Natural Environmental Change
Biodiversity and Conservation
Ecosystems
Environmental Biology
Using Statistics to Understand the
    Environment
Coastal Systems
Environmental Physics
Environmental Chemistry
Biodiversity and Conservation, second
    edition
Ecosystems, second edition

Titles under Series Editor:
David Pepper

Environment and Society texts

Environment and Philosophy
Environment and Social Theory
Energy, Society and Environment,
    second edition
Environment and Tourism
Gender and Environment
Environment and Business
Environment and Politics, second edition
Environment and Law
Environment and Society
Environmental Policy
Representing the Environment
Sustainable Development

Routledge introductions to environment series

Sustainable Development

Susan Baker

 Routledge
Taylor & Francis Group

LONDON AND NEW YORK

First published 2006
by Routledge
2 Park Square, Milton Park, Abingdon, Oxon OX14 4RN

Simultaneously published in the USA and Canada
by Routledge
270 Madison Ave, New York, NY 10016

Reprinted 2006, 2007

Routledge is an imprint of the Taylor & Francis Group, an informa business

Typeset in Times and AG Book by
Keystroke, Jacaranda Lodge, Wolverhampton
Printed and bound in Great Britain by
TJ International Ltd, Padstow, Cornwall

British Library Cataloguing in Publication Data
A catalogue record for this book is available from the British Library

Library of Congress Cataloging in Publication Data
Baker, Susan, 1955–
    Sustainable development / Susan Baker.
      p. cm. – (Routledge introductions to the environment series)
    Includes bibliographical references and index.
      1. Sustainable development. 2. Globalization–Economic aspects.
3. Globalization–Environmental aspects. 4. Economic policy–Environmental
aspects. 5. International organization. I. Title. II. Series: Routledge introductions
to environment series.
HC79.E5B347 2006
338.9'27–dc22   2005015319

ISBN10: 0–415–28210–1   ISBN13: 978–0–415–28210–9 (hbk)
ISBN10: 0–415–28211–X   ISBN13: 978–0–415–28211–6 (pbk)

In memory of my father, Mícheál Bácaoir, for showing me the signs of the ebbing tide and how to wait in silence for the *spideog* to land

Contents

Illustrations

Figures

Table

Series editor's preface

The modern environmentalist movement grew hugely in the last third of the twentieth century. It reflected popular and academic concerns about the local and global degradation of the physical environment which was increasingly being documented by scientists (and which is the subject of the companion series to this, *Environmental Science*). However it soon became clear that reversing such degradation was not merely a technical and managerial matter: merely knowing about environmental problems did not of itself guarantee that governments, businesses or individuals would do anything about them. It is now acknowledged that a critical understanding of socio-economic, political and cultural processes and structures is central in understanding environmental problems and establishing environmentally sustainable development. Hence the maturing of environmentalism has been marked by prolific scholarship in the social sciences and humanities, exploring the complexity of society–environment relationships.

Such scholarship has been reflected in a proliferation of associated courses at undergraduate level. Many are taught within the 'modular' or equivalent organisational frameworks which have been widely adopted in higher education. These frameworks offer the advantages of flexible undergraduate programmes, but they also mean that knowledge may become segmented, and student learning pathways may arrange knowledge segments in a variety of sequences – often reflecting the individual requirements and backgrounds of each student rather than more traditional discipline-bound ways of arranging learning.

The volumes in this *Environment and Society* series of textbooks mirror this higher educational context, increasingly encountered in the early twenty-first century. They provide short, topic-centred texts on social science and humanities subjects relevant to contemporary society–environment relations. Their content and approach reflect the fact that each will be read by students from various disciplinary backgrounds, taking in not only social sciences and humanities but others such as physical and natural sciences. Such a readership is not always familiar with the disciplinary background to a topic, neither are readers necessarily going on to develop further their interest in the topic. Additionally, they cannot all automatically be thought of as having reached a similar stage in their studies – they may be first-, second- or third-year students.

The authors and editors of this series are mainly established teachers in higher education. Finding that more traditional integrated environmental studies and specialised texts do not always meet their own students' requirements, they have often had to write course materials more appropriate to the needs of the flexible undergraduate programme. Many of the volumes in this series represent in modified form the fruits of such labours, which all students can now share.

Much of the integrity and distinctiveness of the *Environment and Society* titles derives from their characteristic approach. To achieve the right mix of flexibility, breadth and depth, each volume is designed to create maximum accessibility to readers from a variety of backgrounds and attainment. Each leads into its topic by giving some necessary basic grounding, and leaves it usually by pointing towards areas for further potential development and study. There is introduction to the real-world context of the text's main topic, and to the basic concepts and questions in social sciences/humanities which are most relevant. At the core of the text is some exploration of the main issues. Although limitations are imposed here by the need to retain a book length and format affordable to students, some care is taken to indicate how the themes and issues presented may become more complicated, and to refer to the cognate issues and concepts that would need to be explored to gain deeper understanding. Annotated reading lists, case studies, overview diagrams, summary charts and self-check questions and exercises are among the pedagogic devices which we try to encourage our authors to use, to maximise the 'student friendliness' of these books.

Hence we hope that these concise volumes provide sufficient depth to maintain the interest of students with relevant backgrounds. At the same time, we try to ensure that they sketch out basic concepts and map their territory in a stimulating and approachable way for students to whom the whole area is new. Hopefully, the list of *Environment and Society* titles will provide modular and other students with an unparalleled range of perspectives on society–environment problems: one which should also be useful to students at both postgraduate and pre-higher education levels.

David Pepper
May 2000

Series International Advisory Board

Australasia: Dr P. Curson and Dr P. Mitchell, Macquarie University

North America: Professor L. Lewis, Clark University; Professor L. Rubinoff, Trent University

Europe: Professor P. Glasbergen, University of Utrecht; Professor van Dam-Mieras, Open University, The Netherlands

Acknowledgements

This book was written while I held a Royal appointment as King Carl XVI Gustaf Professor in Environmental Science. I would like to thank His Majesty King Carl XVI Gustaf of Sweden for providing me with the opportunity to engage in this task. My daughter Niamh deserves special thanks for appreciating how important this engagement was for me and for the understanding she has shown. I would also like to thank the series editor, Dr David Pepper, for his initial suggestion to write this book and for his extremely helpful comments at several points along the way. I am also indebted to the anonymous referees for their most useful comments and observations. I would also like to thank the production team at Routledge, especially Andrew Mould, Zoe Kruze and Anna Somerville, for their patience and understanding as deadlines were set and missed.

The author and publishers would like to thank the following for granting permission to reproduce images in this work: Dr Alan Netherwood, Sustainable Development Unit, Cardiff Council, for the use in Figure 5.1 of the leaflet 'Your World: Our Planet, Our Choice'; the Secretary General of the World Summit on Sustainable Development, United Nations, for use as Figure 3.1 of the Logo from the UN World Summit on Sustainable Development, Johannesburg, 2002; the European Commission for the use as Figure 6.1 of the cover image for *Environment 2010: Our Future, Our Choice: Sixth EU Environment Action Programme*. Every effort has been made to contact copyright holders for their permission to reprint material in this book. The publishers would be grateful to hear from any copyright holder who is not here acknowledged and will undertake to rectify any errors or omissions in future editions of this book.

Susan Baker

Abbreviations

BAT	best available technology
CAP	European Union Common Agricultural Policy
CBD	United Nations Convention on Biological Diversity
CEC	Commission of the European Communities
CFP	European Union Common Fisheries Policy
CITES	Convention on International Trade in Endangered Species of Wild Fauna and Flora
CO_2	carbon dioxide
CoP	Conference of the Parties
CSD	Commission on Sustainable Development
DG	Directorate General of the Commission of the European Communities
EAP	European Union Environmental Action Programme
EEA	European Environment Agency
EU	European Union
FoE	Friends of the Earth
G-77	Group of 77
GDP	gross domestic product
GEF	Global Environment Facility
GMO	genetically modified organism
GNP	gross national product
IBA	important bird area
ICLEI	International Council for Local Environmental Initiatives
IMF	International Monetary Fund
IPCC	International Panel on Climate Change
ISPA	European Union Instrument for Structural Policies for Pre-accession
IUCN	International Union for the Conservation of Nature and Natural Resources
LA21	Local Agenda 21
LDC	least developed country
LIFE	European Union Lending Instrument for the Environment

LULUCF	land use, land-use change and forestry
MEA	multilateral environmental agreement
NGO	non-governmental organization
ODA	official development assistance
OECD	Organization for Economic Cooperation and Development
PHARE	Poland and Hungary Action for Economic Restructuring (Community aid to countries of Central and Eastern Europe)
SAPARD	European Union Special Accession Programme for Agriculture and Rural Development
SIDS	Small Island Developing States
TEN-T	trans-European transport network
TRIPS	Convention and Agreement on Trade-related Aspects of Intellectual Property Rights
UN	United Nations
UNCED	United Nations Conference on Environment and Development
UNDP	United Nations Development Programme
UNEP	United Nations Environment Programme
UNESCO	United Nations Educational, Scientific and Cultural Organization
UNFCCC	United Nations Framework Convention on Climate Change
UNGASS	United Nations General Assembly Special Session
US	United States of America
WBCSD	World Business Council for Sustainable Development
WCED	World Commission on Environment and Development (Brundtland Commission)
WED	women, environment and development
WEDO	Women's Environment and Development Organization
WEHAB	water and sanitation, energy, health, agriculture, biodiversity protection and ecosystem management
WSSD	World Summit on Sustainable Development, Johannesburg, 2002
WTO	World Trade Organization
WWF	the global conservation organization (formerly World Wildlife Fund)

1 Introduction

The environment and sustainable development

Key issues

- **Reconceptualizing development.**
- **Ultimate limits to growth.**
- **Promoting sustainable development; sustainability; the common good.**
- **Three pillars of sustainable development.**

Promoting sustainable development opens up debates about our relationship with the natural world, about what constitutes social progress and about the character of development, both in the North and the South, in the present and into the future. These interrelated issues form the main themes of this book. The book explores the prospects for, and barriers to, the promotion of sustainable development in different socio-economic contexts: the high-consumption societies of the industrialized world, the Third World and the economies in transition in Eastern and Central Europe. The exploration is international in its focus, because it recognizes that promoting sustainable development is a quintessentially global task. In addition, particular efforts to promote sustainable development, including the funding, oversight and physical location of particular projects, usually take place at different locations and across different countries.

Challenging the dominant model of development

The sustainable development model is a challenge to the conventional form of development. Conventional approaches see development as simply modernization of the globe along Western lines. Modernization theory holds that the more structurally specialized and differentiated a society is the more modern and progressive it is (Pepper 1996). To be modernized, a society has to become more technically sophisticated and urbanized and to make increased use of markets for the distribution of economic goods and services. Modernization also brings social changes, including the development of representative democracy, increased mobility and the weakening of traditional elites, kinship groups and communities. Modernization is closely tied to the promotion of individual growth and self-advancement. The transformation of nature, such as taming wilderness into natural parks, harnessing wild rivers to make energy and clearing forests for agricultural production, is one of the hallmarks of modernization.

In the conventional model of development, society is understood to go through different 'stages of economic growth' (Rostow 1960). Traditional societies develop to a stage of economic 'take-off'. With 'take-off', new industries and entrepreneurial classes emerge, as they did in Britain in the nineteenth century. In 'maturity', steady economic growth outstrips population growth. A 'final stage' is reached when high mass consumption allows the emergence of social welfare (Pepper 1996). This model of development assumes a linear progression, in which it becomes necessary for Third World societies to 'catch up' with the Western style of development. This means opening up their economies to Western values, influences and investment and their becoming more integrated into the global market system.

Modern environmentalism has emerged as a critique of this Western-centric development model, although it takes different forms and has different expressions. Environmentalism points to the failure of a model of development that results in unemployment or 'jobless growth' in OECD countries, while the painful transitions in the countries of the former Soviet Union are being accompanied by the tragedies of the failed development strategies for the Third World. Environmentalism challenges many of the basic assumptions that the Western model of development makes about the use of nature and natural resources, the meaning of progress and the ways in which society is governed, including both the traditional patterns of authority within society and how public policy is made and implemented.

Several other social and political movements, such as Marxism and the *dependencia* theories of Third World underdevelopment and dependence, have made similar critiques. However, while environmentalism may make common cause with these arguments, it can be distinguished by its focus on the economic, social *and* ecological dimensions and repercussions of development. Seven key arguments form the backbone of the environmentalist challenge to the Western development model. First, environmentalism takes issue with the understanding of progress found in the Western model. Progress is understood in a limited way, primarily in terms of increased domination over nature and the use of her resources solely for the benefits of humankind. The domination of nature has become a key indicator of human progress (Macnaghten and Urry 1998). Progress is seen, for example, in the clearance of forested land for agricultural production or in the use of natural resources, such as coal, oil and gas, to produce energy in the form of electricity that, in turn, drives production and transport.

Underlying this domination is a reduction of nature merely to a natural resource base, a reduction that values nature only in terms of the use that these resources have for human beings. This gives nature only 'instrumental value', ignoring the 'intrinsic value' of the natural world – that is, the value that nature has over and above its usefulness to humans. Viewing nature instrumentally also leads to neglect of the needs of other, non-human species and life forms.

Second, the Western development model prioritizes economic growth, even though the heightened consumption patterns that it stimulates now threaten the very resource base upon which future development depends. This model assumes environmental deterioration to be an inevitable consequence of development. Although Western society has seen enhanced legal and technical efforts to address environmental pollution, its model of development is none the less premised on the acceptance of a 'trade-off' or exchange between economic development and the environment.

Third, the model assumes that consumption is the most important contributor to human welfare. Here, it is common practice to measure welfare by means of the 'standard of living' – that is, the amount of disposable income that an individual has to purchase goods and services. A development model based on individualistic consumption, rather than fostering social cohesion, leads to increased inequality, especially in an economic system subject to cyclical recession (Ekins 2000). It prioritizes individual self-attainment at the expense of consideration of the common good. In contrast, environmentalism focuses not on the 'standard of living' but on the 'quality of life'. Quality of life refers to the collective, not the individual, level and to enhancing the quality of the public domain, such as through the provision of public education, health care and environmental protection.

Fourth, the model ignores the fact that social stability requires the preservation of natural resources. The deterioration of the natural environment causes social disruption and impairs human health. For example, loss of wild biodiversity in agricultural systems increases the vulnerability of local communities, especially with respect to food supply, which, in turn, leads to social unrest that can undermine social and political institutions (Gowdy 1999).

Fifth, the traditional understanding of development ignores the fact that Western development was, and continues to be, based upon the exploitation not only of the West's own natural resource base but that of many Third World societies, including their timber and ore. The human resources of the Third World have also been exploited. Exploitation has caused underdevelopment in the Third World, not least by creating resource poverty and a culture of dependence. In this environmental view, poverty is caused by the penetration of Western environmentally destructive development models into Third World societies, rather than being alleviated by it. The Western model condemns Third World societies to 'backwardness' while ignoring their long traditions of community resource management. These traditions have developed a body of indigenous knowledge which has enabled many traditional societies to live in harmony with their natural surroundings, although, of course, not all of them have managed to live in this way.

Sixth, the model is blind to the fact that it is not possible to achieve a *global* replication of the resource-intensive, affluent lifestyle of the high-consumption economies of the North. The planet's ecosystem cannot absorb the resultant pollution, as witnessed by climate change. Furthermore, there are not enough natural resources, including water, to support such development. In other words, the model of development pursued by Western industrial societies cannot be carried into the future, either in its present forms or at its present pace.

Finally, and closely related to the previous point, the environmental critique points to the failure of the Western development model to acknowledge that there are limits to economic growth. Limits to growth are imposed by the carrying capacity of the planet, especially the ability of the biosphere to absorb the effects of human activities, and the fact that the amount of resources the planet contains, including water, ore and minerals, is finite. Technological advances, while they may enable society to produce goods with more resource efficiency, will not overcome this limitation. There are thus *ultimate* limits to growth. This means that development has to be structured around the need to adopt lifestyles within the planet's ecological means.

What is significant about this multiple environmental critique of the traditional model of development is that it has shown that the post-war experience of economic growth and prosperity was both exceptional and contingent (Redclift and Woodgate 1997). It was exceptional in that it cannot be replicated across space (from the West to the global level) or across time (into the future). It was contingent upon a short-term perspective, the prioritization of one region of the globe over another, and upon giving preference to one species (humans) over the system as a whole. Environmentalism has also undermined the assumption of a progressive view of society's evolution (Redclift and Woodgate 1997). The environmental critique of development shows that there is no continuous linear development guaranteed for modern society, nor is this development necessarily harmonious (Barry 1999).

The emergence of a new model of development

Many new environmental development models have emerged to replace the old development paradigm. These promote forms of social change that are aimed at fulfilling human material and non-material needs, advancing social equity, expanding organizational effectiveness and building human and technical capacity towards sustainability (Roseland 2000). The objectives of sustainability require the protection of the natural resource base upon which future development depends. For many advocates of the sustainable development model, valuing nature and non-human life forms in an intrinsic way has also to become an integral part of development. The environmental development model is aimed not just at protecting nature, but at creating an ecological society that lives in harmony with nature. This means reconciling economic activity, social progress and environmental protection. In this model, the promotion of human well-being does not have to depend upon the destruction of nature.

The 'sustainable development' model represents an important example of the new environmentalist approach. It seeks to reconcile the ecological, social and economic dimensions of development, now and into the future, and adopts a global perspective in this task. It aims at promoting a form of development that is contained within the ecological carrying capacity of the planet, which is socially just and economically inclusive. It focuses not upon individual advancement but upon protecting the common future of humankind. Put this way, sustainable development would appear to be an aspiration that almost everyone thinks is desirable: indeed, it is difficult *not* to agree with the idea.

Sustainable development is part of new efforts, albeit tentative, to integrate environmental, economic and (more recently) social considerations into a new development paradigm. There are many versions of this new approach. They are united in their belief that there are ultimate, biophysical limits to growth. This challenges industrial societies not only to reduce the resource intensity of production (sustainable production) but to undertake new patterns of consumption that reduce the levels of consumption and change what is consumed and by whom (sustainable consumption). This creates the conditions necessary for ecologically legitimate development, particularly in the Third World.

However, there are many versions of the sustainable development model and not all of them are mutually compatible. There is very little agreement on what sustainable development means and even less agreement on what is required to promote a sustainable future (Redclift and Woodgate 1997).

The Brundtland model of sustainable development

The term 'sustainable development' has been prominent in discussions about environmental policy since the mid-1980s. Following the central role it played in the United Nations (UN) appointed Brundtland Commission (1984–7) and its report, *Our Common Future* (WCED 1987), it has appeared with increasing frequency in academic studies and in government reports. The Brundtland formulation of sustainable development has come to represent mainstream thinking about the relationship between environment and development. It now commands authoritative status, acting as a guiding principle of economic and social development (Lafferty and Meadowcroft 2000).

An increasing number of international organizations and agencies subscribe to at least some, and often most or all, of its objectives (Lafferty and Meadowcroft 2000). These include the European Union (EU), the United Nations Environment Programme (UNEP) and the World Bank. National governments, sub-national regional and local authorities, as well as groups within civil society and economic actors, have all made declaratory and practical commitments to this goal. It also recognizes that promoting sustainable development is a cross-cutting policy task – that is, it cuts across many areas of public policy, including international development, trade, urban and land-use planning, environmental protection, energy policy, agriculture and industry.

The UN has played a particularly prominent role in stimulating engagement with the model of sustainable development. The UN has organized several World Summits, including the United Nations Conference on Environment and Development (UNCED), which took place in June 1992 in Rio de Janeiro, otherwise known as the Rio Earth Summit, and, more recently, the Johannesburg World Summit on Sustainable Development (WSSD), held in 2002. The Rio Declaration, which arose from the Rio Earth Summit, provides an authoritative set of normative principles – that is, principles that deal with moral issues, including gender equality, intra-generational equity (within a generation), inter-generational equity (between generations) and justice. It also details the governance principles needed to deal with how to manage and organize the promotion of sustainable development within society, in institutions and at the political level. This activity has advanced understanding of what sustainable development means. The Summits have also led to several internationally binding environmental agreements, including the UN Framework Convention on Climate Change (UNFCCC) and its related Kyoto Protocol, as well as the Convention on Biological Diversity (CBD). The UN engagement has also led to a proliferation of institutions and organizations, including ones within civil society and from the business community, with a remit to promote sustainable development, such as the Women's Environment and Development Organization (WEDO) and the World Business Council for Sustainable Development (WBCSD). The 'UNCED process' is used as a shorthand way to indicate the range of activities that have taken place under the auspices of the UN since the publication of the Brundtland Report. This range of activities, from the World Summits to the development of legally binding agreements, from the engagement of states to enhancing the role of civil society and business interests is investigated in this book. In addition, the normative and governance principles that have come to be associated with the term 'sustainable development' are explained and explored.

Clarifying the terms used

There is need for some conceptual clarity at this point. This book is about sustainable development. It is not about the concept of sustainability. The term 'sustainability' originally belongs to ecology, and it referred to the potential of an ecosystem to subsist over time (Reboratti 1999). By adding the notion of development to the notion of sustainability, the focus of analysis shifted from that of ecology to that of society. The chief focus of sustainable development is on society, and its aim is to include environmental considerations in the steering of societal change, especially through changes to the way in which the economy functions.

Promoting sustainable development is about steering societal change at the interface between:

- *The social*: this relates to human mores and values, relationships and institutions.
- *The economic*: this concerns the allocation and distribution of scarce resources.
- *The ecological*: this involves the contribution of both the economic and the social and their effect on the environment and its resources.

These are known as the three dimensions or pillars of sustainable development (Ekins 2000).

Sustainable development is a dynamic concept. It is not about society reaching an end state, nor is it about establishing static structures or about identifying fixed qualities of social, economic or political life. It is better to speak about *promoting*,

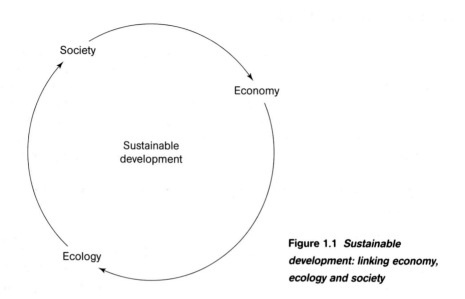

Figure 1.1 *Sustainable development: linking economy, ecology and society*

not achieving, sustainable development. Promoting sustainable development is an on-going process, whose desirable characteristics change over time, across space and location and within different social, political, cultural and historical contexts. Therefore, in this book the expression 'promoting sustainable development' is used to show sensitivity to evolving understandings of, and judgements about, what constitutes sustainable development (sensitive across time) and to the fact that different societies, cultures and groups may aspire to different sustainable development pathways (sensitive across space).

Adopting a dynamic understanding of sustainable development also helps us to recognize that alternative futures lie before society. The promotion of sustainable development is about visioning these alternative futures and, through attitudinal and value changes, policy innovations, political transformations and economic restructuring, embracing a future that is sustainable. While this will be different for different societies, across space and time, there are certain global or common 'baseline' conditions that are required if humanity is to embrace a future that is sustainable. These include a healthy ecosphere and biosphere. They also include adherence to certain normative principles and acceptance of guidelines about what constitutes good governance practice, issues that are given particular attention in this book. This is what is meant by saying that promoting sustainable development requires recognition of the common good. Its challenge is to ensure that society moves along a social trajectory that avoids both the pathways that lead to a direct deterioration of the social state and those that lead to a situation from which further progress is impossible (Meadowcroft 1999).

The governance challenge

Governance can be understood as steering society towards collective goals. However, the environmental challenge to the traditional model of development has led to a questioning of the traditional modes of governance within society and at the international level. Environmentalism has, for example, challenged the ability and legitimacy of traditional forms of government intervention and policy making to address the complex issues posed by the promotion of sustainable development. Rather than being the task of national governments acting alone and using traditional policy means, promoting sustainable development requires engagement across all levels of social organization, from the international, national, sub-national, societal to the level of the individual.

The rise of global environmental problems, such as climate change, biodiversity loss and deforestation, has led to a growing demand for international interventions to deal with both transboundary and global environmental matters. This has stimulated the rapid growth of international environmental laws and management and administration regimes (Gupta 2002). The globalization of environmental governance has been accompanied by pressure to try new and innovative procedures, including expanding the range and role of the non-state actors involved. This includes enhancing the involvement of business interests as well as non-governmental organizations (NGOs) – that is, organizations operating at the national and, increasingly, at the international level, which have administrative structures, budgets and formal members and which are non-profit-making. Thus, while states continue to serve as the primary repositories of authority in relation to environmental management (Young 1997a), there are increased calls for more participatory practices so as to enhance both the legitimacy and democratic nature of the way in which society engages with the promotion of sustainable development. At the same time, sub-national, regional and local engagement also acts as a pressure for the development of new forms of governance, not least so that regional variations, capacity and needs can be taken into account in development plans.

The term 'new practices of environmental governance' refers to the participation of non-state actors, alongside state and international organizations, as well as the utilization of a wide range of policy instruments (including legal, voluntary and market instruments) and normative and governance principles to promote sustainable development.

There is a clear relationship between the type and mode of governance and the success of efforts to promote sustainable development. With this in mind, the book explores whether, and if so to what extent, the commitment of an increasing number of international organizations and agencies to the objectives of the Brundtland model of sustainable development has resulted in changes in the power relations between institutions and societal actors.

The structure of the book

This book is divided into three parts. Part I presents a theoretical and conceptual exploration of sustainable development. Part II looks at the multilevel engagement with sustainable development, including international efforts and the involvement of the sub-national, local level. Part III looks at the promotion of sustainable development in different social, political and economic contexts.

Part I

Chapter 2 provides the conceptual framework that informs the discussions in the rest of the book. The chapter explores the evolution of the meaning and use of the concept of 'sustainable development'. It begins by briefly tracing the development of the concept from its early use in resource ecology to its eventual adoption as a norm of global environmental politics. It explores the variations in meaning and subsequent disputes over the value of the concept, and pays particular attention to the authoritative Brundtland formulation. It also explores the criticism that sustainable development is premised upon a strong 'anthropocentric' approach towards the environment that promotes a managerial relationship with nature. The elaboration of a unifying or precise definition of the concept is less important than understanding the political, economic and social challenges presented by efforts to promote sustainable development in practice. The ladder of sustainable development, as elaborated by Baker *et al.* in 1997, is updated and used to explore the range of normative and governance principles, as well policy issues associated with the promotion of sustainable development at the global level.

Part II

Chapter 3 explores the rationale behind, significance of and theoretical explanations for the construction of a global regime for the promotion of sustainable development. It aims to develop a historically informed, critical awareness of the significant role played by the UN. It looks at how the understanding that the promotion of sustainable development is a global challenge has been stimulated by and, in turn, has stimulated a new era of global environmental governance. Attention is also paid to the association of sustainable development with specific governance principles. The steps are followed from the Brundtland Report, *Our Common Future*, to the Rio Earth Summit, Rio + 5 and onwards to the WSSD held in 2002 and the on-going reportage, monitoring and evaluation regimes established under the auspices of UNCED. This will familiarize the reader with historical developments at the global level and their institutional expressions.

Chapter 4 provides the reader with an understanding of the links between the promotion of sustainable development and the resolution of certain, critical, global environmental problems. The Rio Earth Summit led to two binding conventions, on climate change and on biological diversity. Both conventions are examined in some detail, as they raise a number of key issues that are of concern for this book. In relation to climate change, these include the marked imbalance in resource use between the industrialized world and the Third World, and hence the differences in the burden each is placing on the limited carrying capacity of the environment. It also gives an ideal opportunity to explore the way in which the principles of sustainable development help shape the concrete responses taken to particular environmental problems. In relation to biodiversity, there are growing disputes between the interests of the biotechnology industry of the industrialized world and Third World countries over who should have access to and use of plant and animal genetic resources. Efforts both to manage climate change and to conserve biodiversity throw into sharp relief the tension between economic development and environmental protection, both within the developing world and also within the high-consumption societies of the West.

Chapter 5 explores the tensions involved in global regimes seeking to facilitate bottom-up engagement with sustainable development. The promotion of sustainable development is being encouraged by top-down, global environmental management regimes. At the same time, UNCED is also encouraging bottom-up engagement. Local Agenda 21 (LA21) is the most important action-orientated, bottom-up initiative to emerge from the UNCED process. This chapter begins with an outline of the aims and objectives of LA21. It then goes on to explore the experiences within several countries in organizing LA21. This will include short case studies of countries from both the industrialized world and the Third World. The extent to which LA21 contributes to new forms of participatory governance that help to promote sustainable development, and the structural challenges involved in that undertaking, is critically assessed.

Part III

Chapter 6 outlines efforts to promote sustainable development in high-consumption societies. The main empirical focus in this chapter is on the

European Union. The EU provides an important exemplar of efforts to translate into practice the declaratory statements issued after the Rio Earth Summit. The extent to which EU practice is in keeping with the spirit and principles of Rio is examined. The discussion points to the need for new patterns of sustainable consumption and sustainable production. In the EU context, social actors play a key role in the shift to sustainable consumption; firms and industry, including business interest associations, play a vital role in shifting to more sustainable forms of production. This turns attention to an exploration of the relationship between sustainable development and ecological modernization.

Chapter 7 looks at the promotion of sustainable development in the Third World. The issues raised stand in contrast to the challenges facing high-consumption societies. Protection of the environment and achieving necessary economic development are closely linked with the need to address issues of global justice, poverty and equity in resource use and in the terms of global trade. Both the trade agreements promoted by the World Trade Organization (WTO) and the financial instruments controlled by the World Bank and the Global Environmental Facility are included in the analysis. An additional aim of the chapter is to infuse gender awareness into the study of sustainable development.

Chapter 8 focuses attention on the challenges involved in the countries in transition in Eastern and Central Europe. It asks, within the context of marketization and democratization, what the prospects are for the promotion of sustainable development in transition countries. This question is explored through the lens of the May 2004 Eastern enlargement of the EU.

The conclusion of the book returns to the conceptual and theoretical issues raised in the introduction. Having exposed the reader to detailed and critical discussions of the multi-faceted challenges involved, it asks whether and to what extent the adoption of sustainable development as a norm of global, regional, national and sub-national politics is helping society to find ways in which the tension between economic development and environmental protection can be overcome.

Summary points

- Environmentalism challenges the dominant, Western model of economic development. It argues that this model has a limited understanding of progress, prioritizes growth and fails to recognize the relationship between economic, social and ecological systems.
- The sustainable development model represents a new approach towards development and the steering of social change.
- The Brundtland formulation of sustainable development has attained authoritative status.

Further reading

The Brundtland formulation of sustainable development

Lafferty, W.M. and Meadowcroft, J. (eds) (2000) *Implementing Sustainable Development: Strategies and Initiatives in High Consumption Societies*, Oxford: Oxford University Press.

World Commission on Environment and Development (1987) *Our Common Future*, Oxford: Oxford University Press.

Sustainable development and the environment

Barry, J. (1999) *Environment and Social Theory*, London: Routledge.

Lafferty, W.M. and Langhelle, O. (eds) (1999) *Towards Sustainable Development: On the Goals of Development and the Conditions of Sustainability*, Basingstoke: Macmillan.

O'Riordan, T. (ed.) (2001) *Globalism, Localism and Identity: New Perspectives on the Transition to Sustainability*, London: Earthscan.

Paehlke, R.C. (1989) *Environmentalism and the Future of Progressive Politics*, New Haven, CT: Yale University Press.

Redclift, M. (1987) *Sustainable Development: Exploring the Contradictions*, London: Routledge.

Environmental governance

Gupta, J. (2002) 'Global sustainable development governance: institutional challenges from a theoretical perspective', *International Environmental Agreements: Politics, Law and Economics*, 2: 361–88.

Kooiman, J. (ed.) (1993) *Modern Governance: New Government–Society Interactions*, London: Sage Publications.

Young, O.R. (eds) (1997) *Global Governance: Drawing Insights from the Environmental Experience*, Cambridge, MA: MIT Press.

Part I

Theoretical and conceptual exploration of sustainable development

2 The concept of sustainable development

Key issues

- **Sustainable development; the Brundtland formulation.**
- **Sustainable development as a contested political concept.**
- **Ladder of sustainable development; strong and weak sustainable development.**
- **Normative principles of sustainable development.**
- **Rejection of the principle of sustainable development.**

This chapter explores the Brundtland understanding of sustainable development. It focuses on the Brundtland formulation because it has achieved authoritative status. An increasing number of international organizations and agencies subscribe to at least some, and often most or all, of its objectives (Lafferty and Meadowcroft 2000).

In this chapter the historical origins of the concept of sustainable development are outlined. The Brundtland formulation is then explored in detail. The 'ladder of sustainable development' is used to organize the different interpretations of sustainable development and their policy imperatives. The key normative principles that are associated with the concept are discussed. Finally, attention is turned to the rejection of the model of sustainable development by certain Green theorists and environmental activists.

The chapter forms the basic conceptual building block necessary for understanding the rest of this book. For this reason, it is indicated when discussions in this chapter are relevant to the issues addressed in the other chapters that follow.

Early use of the term 'sustainable development'

Concern about sustainability can be traced back to Malthus (1766–1834) and William Stanley Jevons (1835–82) and other eighteenth- and nineteenth-century thinkers who were worried about resource scarcity, especially in the face of population rise (Malthus) and energy (coal) shortages (Jevons). The issue was raised in the 1950s in the writings of Fairfield Osborn (1953) and Samuel Ordway (1953). It was not until the 1960s and the 1970s, however, that a significant segment of public opinion expressed such unease. These decades were marked by the intensification of anxiety about the environment, particularly the health hazards caused by industrial pollution. This led, in turn, to environmental critiques of conventional, growth-orientated, economic development.

Initially, this concern led to calls, in some quarters, for zero-growth strategies, especially following the publication of the 1972 Club of Rome report, *The Limits to Growth* (Meadows *et al.* 1972). The report, undertaken by a group of young scientists from the Massachusetts Institute of Technology, concluded that, if present trends in population growth, food production, resource use and pollution continued, the carrying capacity of the planet would be exceeded within the next 100 years. The result would be ecosystem collapse, famine and war. The 'limits to growth' argument was also taken up by Herman Daly, who built his 'steady state economics' on the recognition of the absolute limits to economic growth (Daly 1977). However, the 'limits to growth' argument was subject to much criticism. It concentrated only on the physical limits to growth, ignoring the possibility of technological innovations leading to new ways of, for example, addressing pollution or using resources more efficiently in production. It was also seen to present an overly pessimistic view of the rate of resource depletion on a global scale. The argument was displaced by a new belief that environmental protection and economic development could become mutually compatible, not conflicting, objectives of policy. However, this did not necessarily undermine the 'limits to growth' argument, but rather it modified its focus, pointing to the need to limit growth in some areas, to allow for necessary growth in others. This presents the enormous challenge of sorting out when and what type of growth is, or is not, acceptable (Paehlke 2001).

The term 'sustainable development' came into the public arena in 1980 when the International Union for the Conservation of Nature and Natural Resources presented the *World Conservation Strategy* (IUCN 1980). It aimed at achieving sustainable development through the conservation of living resources. However, its focus was rather limited, primarily addressing ecological sustainability, as opposed to linking sustainability to wider social and economic issues.

The Brundtland formulation

It was not until 1987, when the World Commission on Environment and Development (WCED) published its report, *Our Common Future*, that the links between the social, economic and ecological dimensions of development were explicitly addressed (WCED 1987). The WCED was chaired by Gro Harlem Brundtland, the then Norwegian Prime Minister, and *Our Common Future* is sometimes known as the Brundtland Report. The establishment of the WCED and its links with the emerging system of international environmental management, or governance, are discussed more fully in Chapter 3. In this chapter the task is to clarify and discuss the principles, values and norms that have come to be associated with the term 'sustainable development' and the policy imperatives that have become, in turn, linked with them.

The Brundtland Report makes four key links in the economy – society – environment chain (Box 2.1). A good example of environmental linkages is provided when deforestation leads to soil erosion, which, in turn, can cause silting of rivers and lakes. Examples of the linkages between environmental stresses and patterns of economic development are when agricultural policy encourages overuse of chemical fertilizers, which, in turn, can lead to land degradation and water pollution, or when energy policy relies on coal-fired electricity production, which results in greenhouse gas emissions which, in turn, are linked with climate change. Environmental and economic problems are linked with social and political factors, as seen, for example, when rapid population growth leads to stresses on the physical environment. These, in turn, can be related to the position of women in society. Improvements in the social, political, economic and educational position of women in society generally tends to lead to a reduction in the birth rate and a slowdown in population growth. These linkages operate not only within, but also between, nations – many links operate globally. For example, the highly subsidized agriculture of the North erodes the viability of agriculture in the developing countries, as do the terms of international trade. These linkages raise a whole series of issues, which are addressed throughout this book. While Chapter 4 deals with matters related to climate change, the gender dimension is discussed in Chapter 7 and issues of trade and the environment are examined in Chapter 8.

Box 2.1 Causal links in the economy–society–environment chain

- Environmental stresses are linked with one another.
- Environmental stresses and patterns of economic development are linked with one another.

---

continued

> • Environmental and economic problems are linked with social and political factors.
> • These influences operate not only within but also between nations.
>
> *Source:* adapted from WCED (1987: 37–40).

In making the links between the economy, society and the environment, the Brundtland Report puts 'development', a traditional economic and social goal, and 'sustainability', an ecological goal, together to devise a new development model, that of sustainable development. Sustainable development is a model of societal change that, in addition to traditional developmental objectives, has the objective of maintaining ecological sustainability (Lélé 1991). This differs from the previous IUCN approach, mentioned above, which linked the environment with conservation, not with development. In addition, the Brundtland Report made it explicit that social and economic conditions, especially those operating at the international level, influence whether or not the interaction between human beings and nature is sustainable.

The now famous and much popularized Brundtland definition of sustainable development is 'development that meets the needs of the present without compromising the ability of future generations to meet their own needs' (WCED 1987: 43). What is often forgotten is that Brundtland went on to argue that:

> [Sustainable development] contains within it two key concepts: the concept of 'needs', in particular the essential needs of the world's poor, to which priority should be given; and the idea of limitations imposed by the state of technology and social organization on the environment's ability to meet present and future needs.
>
> (WCED 1987: 43)

The Brundtland concept of sustainable development is global in its focus and makes the link between the fulfilment of the needs of the world's poor and the reduction in the wants of the world's rich. It is difficult to distinguish needs from wants, as they are socially and culturally determined. However, in most cultures fundamental needs are similar, and include subsistence, protection, affection, understanding, participation, creation, leisure, identity and freedom (Pepper 1996). The industrialized world consumes in excess of these basic needs, because it understands development primarily in terms of ever increasing material consumption. This excess threatens the planet's ecological resource base and biosystem health. It challenges the industrialized world to keep consumption patterns within the bounds of what is ecologically possible and set at levels to which all can reasonably aspire. This requires changes in the understanding of well-being and what is needed to live a good life. This change allows necessary development in the South. Here economic growth can have, in some contexts, net positive environmental, as well as social and economic, benefits.

> Growth must be revived in developing countries because that is where the links between economic growth, the alleviation of poverty, and environmental conditions operate most directly. Yet developing countries are part of an interdependent world economy: their prospects also depend on the levels and patterns of growth in industrialized nations.
>
> (WCED 1987: 51)

The second focus on limitations, imposed by the state of technology and social organization, presents an optimistic view of our common future. It is optimistic because it presents a vision of the future that contains within it the promise of progress, opened up through technological development and societal change. *Our Common Future* none the less argues that, while technology and social organization can be both managed and improved to make way for a new era of economic growth, limits are still imposed 'by the ability of the biosphere to absorb the effects of human activities' (WCED 1987: 8) and by the need to 'adopt life-styles within the planet's ecological means' (WCED 1987: 9). There are thus ultimate limits to growth. The Brundtland conception of sustainable development does not assume that growth is both possible and desirable in all circumstances.

The idea of ultimate limits is linked with the increasingly popular notion of 'ecosystem health'. Here the health of the ecosystem is the 'bottom line' guiding sustainable development because, if the health of the environment is compromised, everything else is undermined. In this approach, the environment can be seen as a form of 'natural capital' – that is, a resource that can be put to human use (see pp. 33–4). Sustaining this natural capital is a precondition of human life, because 'ecological processes underpin the rest of human activity, and, if these are impaired, then a condition for the very possibility of human activity is impaired too' (Dobson 1998: 44).

There is a danger in this ecosystem approach, however, in that it sustains what is of instrumental value for human beings and does not protect nature for its own sake. Such an approach has strong anthropocentric underpinnings, a matter discussed later in this chapter. A similar anthropocentric approach underlies two other, related concepts. The first, that of 'environmental space', acknowledges that there are limits to the amount of pressure that the earth's ecosystem can handle, without irreversible damage. This leads to a search for the 'threshold level', the level beyond which damage occurs, and the use of this level to set operational boundaries – for example, the level of permitted greenhouse gas emissions. The second is the concept of 'ecological footprint'. Ecological footprint refers to the impact of a community on natural resources and ecosystems, taking account of the land area and the natural capital on which the community draws to sustain its population and production structure (Wackernagel and Rees 1996). The term is particularly useful for looking at the environmental impact of urban development. The more populous and richer a city, the larger its ecological footprint, in terms both of its demands on resources and the size of the area from which those resources are drawn. Many cities not only appropriate resources and carrying capacity from their own rural and resource regions but also from other locations, including globally (Roseland 2000). The concept of ecological footprint has been used in national environmental planning in the Netherlands and forms part of the range of new tools of environmental policy that have developed in recent decades, including environmental assessment and life-cycle analysis (Dresner 2002).

Brundtland hoped that agreement on what type of growth is or is not acceptable, and under what circumstances, could be reached through the development of mutual understanding, through dialogue and through the negotiation of new, and the strengthening of existing, international environmental conventions and agreements. This, in turn, requires new patterns of, and institutions for, global environmental governance – a development discussed in Chapter 3.

The Brundtland formulation presents an optimistic view, especially in relation to the capacity of humankind to engage collectively and constructively in bringing about a sustainable future. It also places strong emphasis on, and hope in, technological development. However, Brundtland envisages building a common future on more fundamental processes of change, which involve not just technological and institutional but also social and economic, as well as cultural and lifestyle changes.

> Sustainable development is a process of change in which the exploitation of resources, the direction of investment, the orientation of technological development, and institutional change are all in harmony and enhance both current and future potential to meet human needs and aspirations.
>
> (WCED 1987: 46)

What is politically significant about the Brundtland Report is that it does not just address the causes of unsustainable development but also puts forward solutions or pathways to the future. This allows the concept to provide a framework for the integration of environmental policies and development strategies into a new development paradigm – one that breaks with the perception that environmental protection can be achieved only at the expense of economic development. The new development paradigm contains many features (Box 2.2).

Box 2.2 The Brundtland development paradigm

Reviving growth

- Changing the quality of growth: making it less material and energy intensive and more equitable in its impact.
- Meeting essential needs for jobs, food, energy, water and sanitation.
- Merging environmental and economic considerations in decision making.

Population and human resources

- Reducing population growth to sustainable levels.
- Stabilizing population size relative to available resources.
- Dealing with demographic problems in the context of poverty elimination and education.

Food security

- Addressing the environmental problems of intensive agriculture.
- Reducing agricultural subsidies and protection in the North.
- Supporting subsistence farmers.
- Linking agricultural production with conservation.
- Shifting the terms of trade in favour of small farmers.
- Addressing inequality in access to and distribution of food.
- Introducing land reform.

Loss of species and genetic resources

- Maintaining biodiversity for moral, ethical, cultural, aesthetic, scientific and medical reasons.
- Halting the destruction of tropical forests.
- Building up a network of protected areas.
- Establishing an international species convention.
- Funding biodiversity preservation.
- Conserving and enhancing the natural resource base.

Energy

- Establishing safe and sustainable energy pathways.
- Providing for substantially increased primary energy use by the Third World.
- Ensuring that economic growth is less energy-intensive.
- Developing alternative energy systems.
- Increasing energy efficiency, including through technological developments and pricing policies.

continued

Industry

- Producing more with less.
- Promoting the ecological modernization of industry.
- Accepting environmental responsibility, especially by transnational corporations.
- Agreeing tighter control over the export of hazardous material and waste.
- Ensuring a continuing flow of wealth from industry to meet essential human needs.
- Reorienting technology and the management of risk.

Human settlement and land use

- Confronting the challenge of urban growth.
- Addressing the problems caused by population shifts from the countryside.
- Developing settlement strategies to guide urbanization.
- Ensuring that urban development is matched by the provision of adequate services.

Source: adapted from WCED (1987).

While the Brundtland model provides a set of guidelines, it is not detailed enough to determine actual policies. These have to be worked out in practice, through, for example, international negotiations. However, as will be seen, a distinction needs to be drawn between what Brundtland argues ought to be the case and what is actually the case in practice, as actors, including governments, at the international, national or sub-national levels, have engaged with the promotion of sustainable development – a gap revealed in several of the following chapters.

Several factors combined to help the Brundtland formulation become the dominant concept in international discussions of the environment and development. First, the formulation offered a way of reconciling what had hitherto appeared to be conflicting societal goals. Second, it came at a time when the problem of environmental deterioration, especially of pollution, was high on the political agenda. This followed the discovery of the ozone hole above Antarctica and the Chernobyl nuclear accident. Third, Brundtland supported developing countries in their pursuit of the goals of economic and social improvement. However, as will become clear throughout this book, many actors, while adopting a commitment to sustainable development, have not embraced the full agenda of change that was envisaged by Brundtland.

Box 2.3 Summary: the Brundtland approach to sustainable development

- It links environmental degradation with economic, social and political factors.
- It presents sustainable development as a model of social change.
- It adopts a global focus.
- It constructs a three-pillar approach: reconciliation of the social, economic and ecological dimensions of change.
- It takes a positive attitude towards development: environmental protection and economic development can be mutually compatible goals and may even support each other.
- It argues that the state of technology and social organization limits development: progress in these areas can open up new development possibilities.
- It recognizes that there are ultimate biophysical limits to growth.
- It takes explicit account of the needs of the poor, especially in the Third World.
- It recognizes that the planetary ecosystem cannot sustain the extension of the high consumption rates enjoyed in industrialized countries upward to the global level.
- It holds that the consumption patterns of the North are driven by wants, not needs. It thus challenges the North to reduce its consumption to within the boundaries set by ecological limits and by considerations of equity and justice.
- It acknowledges the responsibility of present generations to future generations.
- It calls for new models of environmental governance, ranging across all levels, from the local to the global.
- It has achieved authoritative status in international environmental and development discourse and international environmental governance structures and legal frameworks.

Proliferation of terms and meanings

Since the publication of *Our Common Future* there have been numerous attempts to specify exactly what is meant by the term 'sustainable development'. There is a ten-page listing of the most common definitions of 'sustainable development' used in the decade of the 1980s alone (Prezzey 1989; Lélé 1991). It is beyond the scope of this chapter to explore this range.

In addition to the myriad attempts to define sustainable development, the word 'sustainable' has been combined with an array of terms to denote such concepts as 'sustainable growth', 'sustainable cities' and 'sustainable culture'. Not all these applications can be explored either, even though some of them, like the concept 'sustainable growth', are particularly problematic. However, four key expressions in the current discourse on, and debates about, the term 'sustainable development' can be isolated (Box 2.4).

Box 2.4 Key terms in the current discourse on sustainable development

- *Sustainable yield*: maintaining the regenerative capacity of natural systems – for example, forests.
- *Environmental sustainability*: preservation of natural environmental systems and processes, or addressing environmental issues to maintain social institutions and processes.
- *Sustainable society*: living within boundaries established by ecological limits, but linked with ideas of social equity and justice.
- *Sustainable development*: maintaining a positive process of social change.

Source: adapted from Meadowcroft (1999).

A *sustainable yield* is a harvest rate that can, in principle, be maintained indefinitely. The notion of sustainable yield has been widely applied – for example, to fisheries management policy. It has come under particular criticism from environmentalists. They argue that the concept is too narrow in its focus. Its use in calculating sustainability yields from specific fishing stocks ignores the potential for human management to actually disrupt the delicate and poorly understood balances that operate across the marine ecosystem as a whole (Young 2003).

Environmental sustainability is a more ambiguous concept. It can refer to two separate ideas. The first is the sustainability of the processes and systems of the natural environment, such as the climate system or a forest ecosystem. The second is the need to address environmental issues if social institutions and processes are to be maintained (Meadowcroft 1999).

Social sustainability, or the promotion of a sustainable society, refers to a society's ability to maintain, on the one hand, the necessary means of wealth creation to reproduce itself and, on the other, a shared sense of social purpose to foster social integration and cohesion (Ekins 2000). The use of the term, however, has not been straightforward. At the UN Conference on Human Settlements, Habitat II, in Istanbul in 1996, for example, it was not clear whether social sustainability meant the social preconditions for sustainable development or the need to sustain specific social structures and customs (Sachs 1999).

From these examples, it can be seen that the broadening of the concept of *sustainable development*, coupled with its popularity, has given rise to ambiguity and lack of consistency in the use of the term. Some have argued that the concept's ambiguity severely diminishes its usefulness. There is concern among environmentalists that the lack of clarity in the definition allows anything to be claimed as 'sustainable' (Jacobs 1991). It also makes it difficult to devise a set of measurable criteria with which to evaluate whether concrete development programmes are helping to promote sustainable development. Attempts to overcome this problem have led to the elaboration of sustainable development indicators, discussed in Chapter 3.

Sustainable development as a political concept

The proliferation in the meanings of and in the application of the term 'sustainable development' does not necessarily undermine its usefulness. Rather, it reflects the complexity of issues that are invoked when development and environment are juxtaposed (Meadowcroft 1999). As Donella Meadows, one of the authors of *The Limits to Growth*, has argued when discussing some of the linguistic confusion surrounding the use of the word: 'We are struggling for the language now for a whole set of concepts that are urgent in our conversation . . . It's a mess. But social transformations are messy' (quoted in Dresner 2002: 66).

The lack of clarity has also been politically advantageous, because it has allowed groups with different and often conflicting interests to reach some common ground upon which concrete policies can be developed. This is particularly the case within the UNCED process, as discussed in Chapter 3.

More important, the search for a unitary and precise meaning of sustainable development may well rest on a mistaken view of the nature and function of political concepts (Lafferty 1995). As many commentators have argued, sustainable development is best seen as similar to concepts such as 'democracy', 'liberty' and 'social justice' (O'Riordan 1985; Jacobs 1995; Lafferty 1995). For concepts such as these, there is both a readily understood 'first-level meaning' and general political acceptance, but around a given set of core ideas lies a deeper contestation. This makes sustainable development an essentially contested concept (Lafferty 1995). In liberal democracies the debates around these contested concepts form an essential component of the political struggle over the direction of social and economic development – that is, of change (Lafferty 1995). Substantive political arguments are part of the dynamics of democratic politics and the process of conscious steering of societal change. Such arguments are important as they can stimulate creative thinking and practice. One topic on which such creative thinking has occurred is in relation to the idea of 'development', as discussed in the introduction.

The ladder of sustainable development

The diversity of policy options associated with sustainable development can best be seen in terms of a ladder (Table 2.1), originally developed by Baker *et al.* (1997). The ladder offers a useful heuristic device for understanding the variety of policy imperatives that are associated with different approaches to the promotion of sustainable development. These approaches can be adopted by governments, by organizations or by individual Green thinkers or activists. For the purposes of this book, the original version of the ladder has been extensively modified. Its organizing principles have been modified, the rungs of the ladder have been altered and the components of each rung have been made more distinct, and it has been given a more global focus.

Each column in Table 2.1 focuses on a different aspect of sustainable development. Reading across the ladder identifies the political scenarios and policy implications associated with each rung. The ladder also tracks the connection between these positions and particular philosophical beliefs about nature and about the relationship between human beings and the natural world. It helps put flesh on the environmental ethics that underpin practical sustainable development action.

The philosophical underpinning

The varieties of approaches to sustainable development are an indication of differing beliefs about the natural world held in different societies, cultures and historical settings and at the individual level. The values that are attributed to nature range across a broad spectrum, from an 'anthropocentric' to an 'ecocentric' position. At the extreme end of the anthropocentric view, the wealth of nature is seen only in relation to what it can provide for the service of humankind (O'Riordan 1981). In contrast, ecocentrics hold the view that nature has intrinsic value. It is aimed at creating a partnership, based on reciprocity, between human beings and nature.

These two different perspectives have important implications for the design and implementation of policies. The ecocentric approach focuses on the community level and espouses small-scale, locally based technology. The objective is to maintain social and communal well-being and not merely the harmonious use of natural resources (Baker *et al.* 1997). In contrast, the anthropocentric approach can be distinguished by its optimism over the successful manipulation of nature and her resources in the interests and to the benefit of humankind. An extreme example of the anthropocentric approach can be found in the US Wise Use

Movement, a coalition of ultra-conservative politicians, interest groups, scientific institutions and consumers, which promotes economic growth and rejects the need to consider the environment in economic development.

When speaking about sustainable development, making too sharp a distinction between the anthropocentric and ecocentric positions on nature is not wise, however. This is because the main motivation behind any conception or theory of sustainable development is human interest in human welfare (Dobson 1998). This is certainly true of the Brundtland formulation. With its emphasis on human needs, promoting sustainable development is, in this formulation, a way in which to ensure that development (a human activity) is sustainable over time. While this may involve the protection of the natural resource base, the rationale for this protection is essentially a human-centric one: it is protected because it is necessary for our well-being. Nevertheless, ranging attitudes towards nature along a continuum from anthropocentric to ecocentric is useful. At one extreme, nature is seen only in relation to its use to human beings. Moving along the continuum, sustainable development becomes a challenge to devise a more environmentally friendly approach to planning and resource management. Moving further along, nature is allowed to set the parameters of economic behaviour, so that sustainable development becomes an 'externally guided' development model. Reaching the other extreme, so deep is the Green philosophy that sustainable development is viewed as managerial interference with nature and her natural cycles.

Grouping the different policy imperatives in the ladder

At the foot of the ladder is the pollution control approach. It is not that the environment is given no consideration, but rather there is an underlying assumption that, given the freedom to innovate, human ingenuity, especially expressed through technology, can solve any environmental problem (Simon and Kahn 1984). A good example of this approach is found in the so-called Heidelberg Appeal, released by a group of business interests during the Rio Earth Summit. The appeal accepts that environmental protection is an integral part of development, but argues that it should not put limits upon that development, nor should it form our main priority (available at http://www.sepp.org/heidelberg appeal.html, accessed 9 March 2004).

In evidence of the capacity of human beings to manage their environment, supporters of the pollution control approach point to the empirical claim that pollution typically arises in the early stages of industrial development, followed by a stage when pollution is no longer regarded as an acceptable side effect of economic growth and when pollution control policies are introduced (Arrow et al. 1995). There is also a related argument that development follows an

Table 2.1 *The ladder of sustainable development: the global focus*

Model of sustainable development	Normative principles	Type of development	Nature	Spatial focus
Ideal model	Principles take precedence over pragmatic considerations (participation; equity, gender equality, justice; common but differentiated responsibilities)	Right livelihood; meeting needs not wants; biophysical limits guide development	Nature has intrinsic value; no substitution allowed; strict limits on resource use, aided by population reductions	Bioregionalism; extensive local self-sufficiency
Strong sustainable development	Principles enter into international law and into governance arrangements	Changes in patterns and levels of consumption; shift from growth to non-material aspects of development; necessary development in Third World	Maintenance of critical natural capital and biodiversity	Heightened local economic self-sufficiency, promoted in the context of global markets; Green and fair trade
Weak sustainable development	Declaratory commitment to principles stronger than practice	Decoupling; reuse, recycling and repair of consumer goods; product life-cycle management	Substitution of natural capital with human capital; harvesting of biodiversity resources	Initial moves to local economic self-sufficiency; minor initiatives to alleviate the power of global markets
Pollution control	Pragmatic, not principled, approach	Exponential, market-led growth	Resource exploitation; marketization and further closure of the commons; nature has use value	Globalization; shift of production to less regulated locations

Governance	Technology	Policy integration	Policy tools	Civil society – state relationship	Philosophy
Decentralization of political, legal, social and economic institutions	Labour-intensive appropriate, Green technology; new approach to valuing work	Environmental policy integration; principled priority to environment	Internalization of sustainable development norms through on-going socialization, reducing need for tools	Bottom-up community structures and control; equitable participation	Ecocentric
Partnership and shared responsibility across multi-levels of governance (international; national, regional and local); use of good governance principles	Ecological modernization of production; mixed labour- and capital-intensive technology	Integration of environmental considerations at sector level; Green planning and design	Sustainable development indicators; wide range of policy tools; Green accounting	Democratic participation; open dialogue to envisage alternative futures	
Some institutional reform and innovation; move to global regulation	End-of-pipe technical solutions; mixed labour- and capital-intensive technology	Addressing pollution at source; some policy co-ordination across sectors	Environmental indicators; market-led policy tools and voluntary agreements	Top-down initiatives; limited state–civil society dialogue; elite participation	
Command-and-control state-led regulation of pollution	Capital-intensive technology; progressive automation	End-of-pipe approach to pollution management	Conventional accounting	Dialogue between the state and economic interests	Anthropocentric

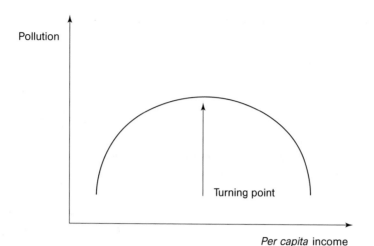

Figure 2.1 *Kuznets curve*

environmental 'Kuznets curve' – that is, that pollution starts out low, increases at the early stages of development, but then diminishes as the economy shifts into a less resource-intensive, post-industrial stage.

However, this theory ignores the fact that high-pollution activities can be displaced from the industrialized to the developing world, thereby reducing pollution in one place but not at the overall, global level. Japan provides a good example of such behaviour, as it has the aluminium needed for its industrial production smelted elsewhere and uses the forest resources of other countries to provide the packing needed for the consumer goods it produces, while maintaining high levels of forest protection at home.

Immediately above the pollution control approach on the ladder is the concept of 'weak' sustainable development, whose aim is to integrate capitalist growth with environmental concerns. This position is closely associated with David Pearce and the highly influential *Blueprints* for a green economy reports (Pearce *et al.* 1989; Pearce 1994, 1995; Pearce and Barbier 2000). These argue that the best way to preserve critical 'natural capital', which comprises important natural resources or processes such as forests or the climate system (see p. 33), is to give it an economic value or price. The price is based on what people would be willing to pay to protect that natural capital. This figure can then be used to undertake a 'cost–benefit analysis', which works out the gains and losses involved in using the natural capital. If the gains outweigh the losses, then the natural capital should be used, or 'drawn down'. However, this idea of 'putting a price on the planet' has been severely criticized (Dresner 2002). The anthropocentric basis of this position is clear, and some argue that much of nature is 'beyond price'. It also breaches many of the normative principles that have come to be associated with sustainable development:

> Pearce's work shows quite well the way in which the application of cost–benefit analysis to global environmental issues works against the principles of intergenerational equity and intragenerational equity that lay at the core of the Brundtland Commission's definition of sustainable development. Because decisions are based on ability to pay, less weight is given to the interests of the poor and the future. . . . It is hard to see how [money] is a good measuring rod when comparing the preferences of Americans and Bangladeshis, or people today and people a hundred years from now.
>
> (Dresner 2002: 116)

The object of policies to promote weak sustainable development remains economic growth, but environmental costs are taken into consideration through, for example, accounting procedures. This is possible because the environment is considered to be a measurable resource. Weak sustainable development has a growing influence on international agencies, including the World Bank, discussed in Chapter 7. It has led to the development of environmental management and of many new environmental policy tools, including Environmental Impact Assessment, and of adjustments to the market to take account of market failure through fees, taxes and tradable permits.

On the third rung of the ladder is the 'strong' sustainable development position. Whereas Pearce asserts that economic development is a precondition for environmental protection, strong sustainable development asserts that environmental protection is a precondition for economic development (Baker *et al.* 1997). One of the major differences between strong and weak forms of sustainable development is in relation to whether 'natural capital' (oil, for example) can be drawn down and technology can be used as a substitute (replacing oil by solar technology), or whether there is such as thing as 'critical' natural capital, which cannot be replaced by technology and should be preserved absolutely. This disagreement is also about how development is structured: should the proceeds from running down natural resources, such as the UK's North Sea gas, be invested in the development of new technologies that can replace these resources? Or should they be invested in other forms of capital, such as human capital, through investment in education (Dresner 2002)? Weak sustainability assumes almost total substitutability by technology, whereas strong sustainability assumes some substitutability but imposes strict limits on how much human capital can compensate for running down natural capital.

Within strong sustainable development, there is also a tendency to hold that, given the limited understanding of the complexity of the natural environment, the precautionary principle should be adopted, especially in relation to the management of risk, discussed in Chapter 4. The 'precautionary principle' holds that, in the face of risk and of uncertain scientific knowledge, policy makers should err on the side of caution. For example, while there may be uncertainties about the science, climate change brings high economic, social and ecological risks, so governments should take action now, lest it be too late to address the problem adequately in years to come. Promoting this form of sustainable development requires strong state intervention (government) to be combined with new forms of participation (governance). For example, governments need to ensure adequate market regulation and develop new energy and transport policies to deal with climate change. The involvement of consumers, economic interests and local communities is also needed to bring about changes in consumption patterns and to ensure that society makes more use of environmentally friendly transport modes. Thus the stronger form of sustainable development does not give market forces free rein to determine behaviour.

Strong sustainable development also seeks a shift from quantitative growth, where growth is seen as an end in itself and a measure only in material terms, to qualitative development, where quality of life is prioritized. This has led several Green economists to develop alternative indicators for the traditional GNP measure of human welfare. Herman Daly's Index of Sustainable Economic Welfare is among the best known of these. It includes calculations for depletion of natural capital, the cost of pollution and social issues such as unemployment and inequality (Daly and Cobb 1990). This position can be said to promote sustainable development only in so far as it rejects the overall objective of economic growth. The weak form of sustainable development cannot perpetuate itself indefinitely, as it permits the draw-down of natural resources in order to support production. The stronger form of sustainable development permits growth only under certain limited conditions: when it is designed to deal with necessary development, as in the Third World, and when it is balanced by reduction in growth elsewhere. Strong forms view development over the long term and at the global level.

The top rung of the ladder represents the ideal approach to sustainable development. It offers a profounder vision aimed at structural change in society, the economy and political systems. Some who hold this position reject the idea of sustainable development *as formulated by Brundtland*, but have gone on to modify the Brundtland position by injecting it with more radical, socialist considerations (Pepper 1998).

For others, severe restraints on the consumption of the Earth's resources and on humankind's related economic activities and, most controversially, a reduction in human population, are all proposed. Here promoting sustainable development is premised upon a radical change in our attitude to nature. Perhaps the most radical of these deep Green positions is that of 'deep ecology', originally put forward by the Norwegian philosopher Arne Naess. The deep ecology position has three characteristics: attributing equal value to all life forms; seeking identification with non-human natural entities and systems; and advocating the development of policies that stress non-interference and the harmony of human life and nature (Naess 1989). The substitution of human for natural capital is not allowed and gains in human welfare at the expense of radical transformation of the 'natural' environment are not tolerated (Sylvan and Bennett 1994). Within deep ecology, there is a rejection of the assumption that human beings can and ought to manage the environment (Katz *et al.* 2000). As the management of the environment, albeit to different degrees, is an underlying assumption of all efforts to promote sustainable development, it thus rejects the sustainable development project. The environmental groups Earth First!, Sea Shepherd and the Animal Liberation Front represent examples of how deep Green philosophy can influence environmental direct action.

So far, in the discussion of the ladder of sustainable development, the different approaches towards nature and the natural world have been sketched along a continuum. The sustainable development project is rejected at either extreme, but for exactly opposite reasons: for the pollution control approach, promoting sustainable development is seen as threatening economic growth by taking environmental considerations *too much* into account; at the opposite extreme, deep ecologists argue that sustainable development displaces considerations of nature, thus taking the environment *too little* into account (Jagers 2002). Occupying the middle ground are a range of different understandings of sustainable development; each in turn can be associated with different policy imperatives. Implicit in the discussion so far has been that there is a link between the promotion of sustainable development and certain ideas about what constitutes right conduct (morals). Mention has been made, for example, of links with issues of justice and equity. These moral ideas have begun to permeate the discourse on sustainable development and, as a result, the promotion of sustainable development has now come to be associated with certain norms, or authoritative standards, of behaviour. To complete the discussion of the concept of sustainable development, these normative principles are explored in greater depth.

The normative principles of sustainable development

As international engagement with the concept of sustainable development progressed onwards from the time of the Brundtland Report, the term began to be associated with a number of normative principles. Normative principles are moral statements that specify what is good or bad, and mould attitudes and guide behaviour. In Brundtland, they were primarily associated with meeting human needs, especially the development needs of the poor and the protection of environmental resources, including global environmental systems such as the climate system. However, Brundtland also introduced other normative aspects into the discussion, and this opened the way for a range of normative principles to come to be associated with the term (Box 2.5).

Box 2.5 Normative principles of sustainable development

- Common but differentiated responsibilities.
- Inter-generational equity.
- Intra-generational equity.
- Justice.
- Participation.
- Gender equality.

Common but differentiated responsibilities

The 1992 Stockholm Declaration (see Chapter 3) proclaimed the responsibility of governments to protect and improve the environment for present and future generations. After the Stockholm Conference, several states recognized in their constitutions or laws the right of their citizens to an adequate environment as well as the obligation of the state to protect the environment. This idea of environmental responsibility was further elaborated by the Brundtland Report, which called upon all governments to take responsibility for the environment, because the promotion of sustainable development involves guarding the *common fate* of humanity. However, in pursuing this responsibility, account has to be taken of the fact that not all countries have contributed in the same way, or to the same extent, to the current environmental crisis. Moreover, countries have different capacities to take effective action to deal with, or prevent further, environmental deterioration. This principle of 'common but differentiated responsibilities' provides a way of distributing the responsibilities and tasks associated with the promotion of sustainable development more fairly among the Third World and industrial countries. The principle acknowledges that industrial countries have been the main contributor to environmental problems through their patterns of resource exploitation, production and high consumption. It also recognizes the unequally borne economic effects of implementing international environmental laws and agreements. Further, it takes account of the different capacities, including financial and technical capacities, available within countries to address the problem. In short, the use of the principle of common but differentiated responsibilities is driven by equity considerations.

Subsequently, the principle was to influence the Rio Declaration, the agreement that was concluded at the Rio Earth Summit in 1992, as discussed in Chapter 3. Principle 7 of the Rio Declaration reads: 'In view of the different contributions to global environmental degradation, States have common but differentiated responsibilities.'

The Rio Declaration called for the diffusion of *differential obligations* through international environmental law as a way of putting the principle into practice. This was not an entirely new development. Prior to the Rio Earth Summit, differential obligations had already appeared in several international legal conventions and agreements, such as the Convention on the Long-range Transport of Air Pollution (1978) and the Montreal Protocol on Ozone Depleting Substances (1988). The use of differential obligations was especially marked during the 1990s.

Differential obligations can take several forms. It can oblige industrialized countries, for example, to transfer technology or resources, including finance and expertise, to developing countries. It can also lead to variations in the quantified targets and goals set for different countries. The Kyoto Protocol, an international agreement designed to address climate change, obliges industrialized countries to reduce their carbon emissions according to a negotiated scale, while developing countries have, as yet, no reduction targets to meet. Countries can also be given different time scales for implementing their obligations, as well as compensation, funding and resources to help with implementation. Differential obligations can last indefinitely or can expire. The negotiations leading up to the second commitment period of the Kyoto Protocol (starting in 2012), for example, will see a review of current agreements.

While drawing upon considerations of equity, the use of the principle of common but differentiated responsibilities has a strong functional logic. It is often used as a means of ensuring that developing countries sign up to and continue to participate in international environmental management regimes, such as the climate change regime mentioned above. Developing countries may be more motivated to implement conventions that acknowledge their vulnerability in the face of an environmental crisis that they did not primarily cause. When international conventions are faced with pervasive, multi-causal problems that traverse national boundaries, such participation is highly valued (Iles 2003). The use of the principle thus helps to ensure that efforts to promote sustainable development have a more global reach.

Despite its widespread use, the principle continues to be contested in international environmental negotiations. Acrimonious debates also took place over whether the principle should be included in the Johannesburg Declaration released at the end of the 2002 World Summit on Sustainable Development (WSSD), discussed in Chapter 3. The fact that the United States failed in its repeated attempts to have the phrase omitted from the declaration has been portrayed as perhaps the greatest achievement of the WSSD (Iles 2003).

There is nevertheless a problematic side to the use of the principle of common but differentiated responsibilities. Developing countries, particularly during negotiations on international environmental treaties, often argue that environmental protection measures can interfere with their economic development strategies. The use of the principle may thus help perpetuate the perception that a trade-off exists between the environment and development, despite the fact that the model of sustainable development is designed to break such a perception. There is also the possibility that differential obligations can reinforce environmental degradation by permitting Third World countries to continue polluting, destroying habitats or overusing their resources. The irony is that this behaviour can run down a county's natural resource base and destroy livelihoods, thus helping to perpetuate the very differences that were used to justify a country's special treatment in the first place (Iles 2003). Thus care needs to be taken in the way in which the principle is put into practice if it is to help promote sustainable development at the global level.

Inter-generational and intra-generational equity

According to the Brundtland Report, the promotion of sustainable development implies concern for both intra-generational and inter-generational equity, especially with respect to resource use:

- Intra-generational equity: refers to equity *within* our own generation.
- Inter-generational equity: refers to equity *between* generations, that is, including the needs of future generations in the design and implementation of current policies.

(WCED 1987: 5–6)

Intra-generational equity

This highlights the importance of meeting the basic needs of present generations, given the widely uneven pattern of global development. The notion of equity within generations owes much to the work of John Rawls's *A Theory of Justice*

(1971), although Rawls's work pre-dates current concerns about the global environment. The principle of equity is fundamental to Rawls's theory of justice, in which he argues for equality in the distribution of basic social goods, such as liberty and opportunity, income and wealth and social respect.

Present concern about equity acknowledges the inequity in resource use between the North and the South, the rich and the poor, while at the same time seeing poverty as both a cause and a consequence of unsustainable behaviour. Poverty can lead to the over-exploitation of the resources of a local environment to satisfy immediate needs. 'Those who are poor and hungry will often destroy their imme-diate environment in order to survive' (WCED 1987: 28). Poverty, caused by the failure to address land reform, for example, can lead landless farmers to use ecologically harmful 'slash and burn' agricultural techniques, as is happening in the Amazon rain forest. Poverty can also lead to the growth of urban slums, which lack adequate infrastructure, especially for sewage and waste disposal, resulting in both health and environmental hazards. There is thus a relationship between poverty and exposure to the negative consequences of environmental degradation, such as polluted water. Such concerns led to the development of the environmental justice movement, particularly in the US. This movement primarily addresses the negative impacts of environmental degradation on human health (Martinez-Alier 1999). The environmental justice movement is typified by the actions of a local community in Love Canal in the US in the 1970s, where low-income housing was build on a toxic dump site, which subsequently led to severe health effects in the local community (Merchant 1992). The concerns of the environmental justice movement, however, are narrower than those raised in the sustainable develop-ment agenda. The broader remit of the latter encompasses issues not just of health, but of environmental protection, and the maintenance of biodiversity as well as issues of global equity and justice of access to, and use of, resources.

This links the promotion of sustainable development with questions of power and the removal of the disparities in economic and political relationships between the North and South. For Brundtland, there is a strong functional relationship between social justice and sustainable development, because poverty is a major cause of environmental deterioration and the reduction in poverty is a precondition for environmentally sound development (WCED 1987).

> Developing countries must operate in a world in which the resource gap between developing and industrial nations is widening, in which the industrial world dominates in the rule making of some key international bodies, and in which the industrial world has already used much of the planet's ecological capital. This inequality is the planet's main 'environmental' problem; it is also its main development problem.
>
> (WCED 1987: 46)

However, Brundtland gives priority to the world's poor, independent of any poverty–environment relationship. This is because poverty is seen 'as an evil in itself' and sustainable development requires meeting the basic needs of all, thus extending to all the opportunity to fulfil their aspirations to a better life. The relationship between social and economic justice and physical sustainability is not just functional – that is, it does not merely serve a particular practical and efficiency purpose – but it is also normative – that is, it is based upon ethical considerations (Langhelle 2000).

Making the link between poverty and environmental harm is not to deny that many Third World communities have devised sustainable coping strategies to deal with resource use problems. In addition, it is not only the poor who overuse environmental resources but the rich as well, so that the alleviation of poverty does not necessarily lead to the end of environmental degradation. Poverty relief needs to be combined with other policies if environmental degradation is to be halted (Dobson 1998).

Inter-generational equity

The idea of inter-generational equity dates as far back as the political philosophy of Immanuel Kant (1724–1804), who developed the idea of posterity benefiting from the work of its ancestors. The philosopher Edmund Burke (1729–99) also wrote about the idea of inter-generational partnership (Ball 2000). However, in considering future generations, Brundtland developed a perspective quite the opposite of previous thinkers, arguing that we borrow environmental capital from future generations and that 'Our children will inherit the losses' that this brings (WCED 1987: 8). Brundtland argued that today's society might compromise, in many different ways, the ability of future generations to meet their essential needs (WCED 1987). Rather than focusing upon the ways in which the actions of the present generation may help those of the future, Brundtland focused upon how today's unsustainable behaviour can narrow the options available for future generations. Promoting sustainable development requires foreclosing as few future options as possible (WCED 1987: 46). Green theorists have developed Brundtland's ideas further, to suggest that our relation with other generations creates obligations. This poses a problem, however, as it is unclear how far into the future these obligations stretch. It would seem insufficient to restrict concern to the next generation only, as many environmental problems or processes work on a very long-term scale. A 'glacial' time scale, for example, applies to radioactive waste. Considerations of inter-generational equity also raise another very difficult political issue, namely how future generations can be given some form of voice or consideration in present policy making. Not least among these problems is how to find out what the interests or needs of future generations will be. In addition, environmental management tasks, such as planning, monitoring and evaluation, typically do not fit in with the longer-term period needed to take account of future generations. As such, considerations of inter-generational equity require considerable extension of the time scale of current planning and policy-making models and practices.

Other theorists have argued that the principle of inter-generational equity brings with it more stringent requirements. Dobson in particular has argued that the principle means that future generations' human needs have to take precedence over the present generation's human wants. He argues that it:

> would be odd for those who argue for the sustaining of ecological processes to put the wants of the present generation of human beings (which might threaten those processes) ahead of the needs of future generations of human beings (who depend upon them).
>
> (Dobson 1998: 46)

It has also been argued that once the interests of future generations are taken into account, then concern for many features and aspects of the *non-human* natural world can be generated. This would include concern for other species, which may be essential prerequisites for future generations to meet their needs.

The principle of participation

Rationale

There are both normative and functional reasons why participation is an essential condition for the promotion of sustainable development. Taking a normative perspective, it can be argued that participation in decisions that shape one's life is considered a hallmark of democratic practice. Promoting sustainable development involves making difficult decisions about, for example, reducing consumption levels, or introducing taxes on goods that have a negative environmental impact, or prohibiting or placing restrictions on certain forms of behaviour, such as on the ways of disposing of household waste. It is only through increased participation that society can construct 'a shared public basis' on which to ground the legitimacy and acceptance of such restrictions and corrections (Achterberg 1993). As Brundtland has argued: 'Making the difficult choices involved in achieving sustainable development will depend on the widespread support and involvement of informed public and non-governmental organisations, the scientific community, and industry' (WCED 1987: 21).

A second normative argument is that participation is necessary because promoting sustainable development raises issues of an essentially 'subjective and value-laden' character (Paehlke 1996). These include the value one attributes to nature, for example, whether one wishes to promote nature preservation, or conservation or the unrestricted use of natural resources. Given these value differences, agreement on the objectives of policy is unlikely unless these objectives are reached through participatory practices.

The functional arguments build upon Brundtland's belief that 'effective partici-pation in decision-making processes by local communities can help them articulate and effectively enforce their common interest' (WCED 1987: 47). To this functional reason is added the argument that participation is the only approach to policy making that can incorporate the needs of all segments of society, future generations and other species (Dryzek 1992; Pepper 1998). This claim is some-what contentious, however, and is discussed below when some of the problems associated with participation are explored. Participation is also seen as leading to better social choices (Fiorino 1996), because it is seen as increasing the evidence base for decisions.

Participation, grounded on both normative and functional rationales, is important because it helps deal with the fact that, in liberal societies, reasonable people disagree about ideals or values. If public authorities do not allow explicit dis-cussion about these competing ideals, they undermine the legitimacy of the process of policy making, foster misunderstanding and lead to unwillingness to abide by policy decisions (Bell 2004). From the point of view of promoting sustainable development, participation helps society make decisions about the difficult issue of 'what' is to be sustained and for whom.

The importance of participation was recognized at the Rio Earth Summit and was made most explicit in Agenda 21. Since the Brundtland Report was pub-lished, the environmental movement has been single-minded in seeking more direct involvement of the public in governmental decision making regarding the environment (Paehlke 1996). Indeed, participation is seen as a defining characteristic of sustainable development. A statement by World Humanity Action Trust, an NGO affiliated to the UN, typifies this position, arguing that action by governments alone will not solve the problems underlying the global failure to implement sustainable development. The Trust also believes that, in order to transcend political conflicts and vested interests, multi-stakeholder participation and partnerships need to be established and developed in decision making and implementation. For this group, local and national participation remains at the heart of an integrated policy for the implementation of sustainable development (World Humanity Action Trust 2001).

Types of participation

Participation typically refers to the involvement of those outside the formal governmental apparatus in decision-making processes aimed at making public policy. However, beyond this weak understanding, there is little agreement on what participation actually means (Fiorino 1996). However, a useful distinction can be drawn between 'elite' and 'democratic' forms of participation. Elite participation refers to either expert or interest-group participation in policy making. This form of participation is normal in modern liberal democracies. Many environmental groups and Green political theorists are not content with elite participation, seeking instead another form, which is referred to as democratic participation. This is where people take part in policy making as citizens, not as experts or interest advocates. Here 'participation' means the direct participation of amateurs in public policy making, allowing citizens to participate with administrators and experts on a more equal basis, creating structures for face-to-face interaction over time, and allowing citizens a share in decision making (Fiorino 1996). Mechanisms for this participation would include, at a minimum, the development, and increased use, of hearings and fora for public discussion, right-to-know laws, public inquiries, citizens' groups and town meetings.

> Civil society . . . is a convenient short-hand term to refer to the organizations of non-profit interest groups, which form to assert interests and causes outside state-based and controlled political institutions, which constitute networks of action and knowledge.
>
> (Curtin 1999: 446)

However, not all countries have developed civil societies and the ability of citizens to participate effectively in policy making varies widely. In Chapter 5 examples are discussed of where civil society remains underdeveloped and where there have been few chances to participate in environmental policy-making processes. Similarly, Chapter 8 examines how the weak nature of civil society in the transition states of the former communist bloc acts as a barrier to the promotion of sustainable development.

The call for enhanced citizen participation is closely linked with a 'deliberative' conception of democracy. This conception stresses the importance of on-going dialogue between citizens. This contrasts with the more traditional forms of sporadic, passive, procedural participation, such as voting. The former is seen as the key to democratizing decision-making processes, because it requires greater transparency – that is, citizens have access to information held by public authorities. This form of participation is not aimed at giving a voice to individual preferences or interests for their own sake. Rather, it aims at finding a voice for the common good (Curtin 1999). This is why it is important to the promotion of sustainable development. Enhancing participation requires society to find ways of educating people to participate and to develop ways of reaching agreement on what constitutes the common good. Participating groups may be different in size, stability, financial strength and organizational strength – in short, participation may not be based on equality. This also challenges society to find ways in which this inequality can be addressed.

Participation is not just a means of legitimizing existing sustainable development policies. Rather, it is a necessary part of the formulation, implementation and evaluation of such policies (Lélé 1991). This means that the call for greater participation is also a rejection of the traditional way in which policy is made, a way that relies on technically based decision making aimed at dealing with a 'public interest', even though the public were rarely consulted on what that interest may be. Thus the call for participation is not just a plea for increased public discourse within the context of existing political and administrative structures and constraints, nor is it merely about the utilization of existing institutions and structures of the state and of public administration. For many, it is about creating new structures and new processes for governing society. In this argument, participation is a route to achieving new ways of governing society, not merely an end goal in itself. These new forms of governance are the focus of attention in Chapter 3.

Problems with participation

It is important not to assume that the involvement of civil society, including local and environmental interests, in policy formulation and implementation will necessarily ensure the promotion of sustainable development. In much of the discussion about participation there is a tendency to assume a 'natural' congruence between making decision procedures more open and democratic and sound substantive environmental policy outcomes. However, the assumption that democracy and enhanced environmental protection are mutually reinforcing is open to question (Lafferty and Meadowcroft 1996a). There are good reasons for believing that the relationship between democracy and good environmental practice is far from straightforward. People's understanding of ecological issues, for example, may not extend beyond the effects these have on their own immediate interests (Hayward 1995), or they may not take account of future generations. The rise of Nimbyism (Not in my back yard-ism) provides a good example. Nimbyism is a disparaging term used to describe those who participate in policy making to protect their own narrow interests. This can include, for example, objecting to a particular road development scheme or factory location because it can threaten the value of their property. Furthermore, even when the public sphere has been 'reinvigorated', there is no guarantee that the free and equal conversation will grant a more valued status to the non-human world than it has at present (Dobson 1993).

Participation, for example, opens up the potential for demagogic behaviour and political extremism. This problem does not just operate at the level of the individual. The development strategy of one region, for example, may deprive another of a resource on which its prosperity has traditionally depended. A specific locality may refuse to accept the consequences of policies that the country's citizens collectively endorse. Likewise, a region may wish to accept levels of environmental risk that its neighbours or national authorities find unacceptable (Lafferty and Meadowcroft 1996b).

Participation also throws up another problem – *who* decides what groups or individuals participate and on what basis they participate in policy making? What happens when participating groups hold conflicting views about sustainable development? In such circumstances, should consultation rights be given to avowedly anti-environmental interests? Opening up policy making to groups, including environmental organizations, also means setting up new arrangements that bypass conventional democratic institutions and processes, including processes of accountability and control. This raises the thorny issue of whether or not these new arrangements have democratic legitimacy. Decision making at all levels is allegedly something that happens through elected representatives and through assemblies. However, many of the groups that seek access to the policy-making process are not representative, nor are they accountable. To ensure that participation and the new governance models it is promoting remain accountable and democratic, such arrangements have to be tightly linked into the democratic systems of government, by, for example, making sure that their influence is secured through the formal political process (Blowers 1997). There is still need for action to be backed by legal authority and, if necessary, coercive sanctions (Lafferty and Meadowcroft 1996b).

Gender equality

Prompting sustainable development without consideration of the needs of the female half of the world's population is an empty gesture (Dobson 1996). At a minimum, it breaches the principles of inter-generational and intra-generational equity. This means that account has to be taken of the fact that environmental degradation affects men and women differently. This arises from the different societal tasks men and women have, from their different roles in relation to reproduction and from the differences in access to and distribution of power.

Equitable participation of women in environmental decision making is also a minimum requirement for the promotion of sustainable development. This opens the space for a female-sensitive identification of needs. In addition, by drawing upon the insights, experience and knowledge that women can bring to the problem, it can help to identify a wider range of policy solutions. The links between women and the environment are explored by feminist environmentalism, a position discussed in Chapter 7. Attention is also paid to a somewhat different position, namely the argument that women and men differ in their relation to nature, in their historical contribution to the environmental crisis and in the type of responsibility they have for overcoming the legacy of past behaviour. This ecofeminist argument is also discussed in more detail in Chapter 7.

Conclusion

The popularization of the discourse on sustainable development following the publication of the Brundtland Report should be understood in the broader context of growing ecological awareness, widening disparities between the North and the South, especially the debt burden that has trapped many developing countries in poverty, and the rising opposition to the negative consequences of economic growth in some advanced capitalist nations. Because it offers a way of reconciling economic development and ecological protection, while being flexible enough to allow governments to take account of different political cultures, policy contexts and socio-economic needs, the Brundtland formulation of sustainable development has been able to become *the* guiding principle of international environmental negotiation and governance practice. However, not all environmentalists have endorsed the concept of sustainable development. Some reject the project on philosophical grounds, claiming that its underlying motives are too anthropocentric; by others, it is rejected on political grounds, either because it is not radical enough, or because it is far too radical altogether.

The Brundtland approach is built upon a belief in the common heritage of humankind, trust in our technology, and optimism about our willingness to engage collectively in the protection of our common future. The normative principles that have come to be associated with sustainable development have led to the elaboration of specific rights and obligations for states, and they have acted as guidelines for international and national environmental regulations and laws. These normative principles have widened the scope of those to whom environmental obligations are owed beyond states and beyond present generations. They also place obligations upon the individual, especially as a consumer.

These normative dimensions stretch their demands into the policy-making system, or system of governance, to require that the policy-making processes become more inclusive and gender-sensitive, and facilitate the fuller participation of societal actors in decision making that affects their future. This points to the task of the next chapter, which is to discuss the governance dimensions of sustainable development.

Summary points

- Sustainable development is about the long-term transformation of basic aspects of the present industrial economic system. Promoting sustainable development is about the construction of a new development paradigm, framed within the ecological limits of the planet.
- There has been a proliferation in the meanings and applications of the term, making the search for a precise definition a frustrating effort. More important, it is also a mistaken endeavour in that it misunderstands the function of political concepts.
- The policy imperatives associated with promoting sustainable development can be seen in terms of a ladder, ranging from a weak to an ideal form.
- At either end of the continuum, the sustainable development project is rejected, but for entirely opposite reasons.
- Sustainable development has come to be associated with several normative principles that now guide environmental management practices and international law, but increasingly stretch into other issue areas.
- Promoting sustainable development also requires new governance practices.

Further reading

Basic reports

World Commission on Environment and Development (1987) *Our Common Future*, Oxford: Oxford University Press.

Commentaries on the concept of sustainable development

Baker, S., Kousis, M., Richardson, D. and Young, S. (eds) (1997) *The Politics of Sustainable Development: Theory, Policy and Practice within the European Union*, London: Routledge.

Lélé, S. (1991) 'Sustainable development: a critical review', *World Development*, 19: 607–21.

Radical critiques of sustainable development

The Ecologist (1993) *Whose Common Future? Reclaiming the Commons*, London: Earthscan.

Pepper, D. (1993) *Eco-Socialism: From Deep Ecology to Social Justice*, London: Routledge.

Pepper, D. (1998) 'Sustainable development and ecological modernization: a radical homocentric perspective', *Sustainable Development*, 6: 1–7.

Redclift, M. (1987) *Sustainable Development: Exploring the Contradictions*, London: Routledge.

Sachs, W. (1997) 'Sustainable development', in M. Redclift and G. Woodgate (eds) *The International Handbook of Environmental Sociology*, Cheltenham: Edward Elgar, 71–82.

The normative principles

Dobson, A. (1998) *Justice and the Environment: Conceptions of Environmental Sustainability and Dimensions of Social Justice*, Oxford: Oxford University Press.

Iles, A. (2003), 'Rethinking differential obligations: equity under the Biodiversity Convention', *Leiden Journal of International Law*, 16: 217–51.

Lafferty, W.M. and Meadowcroft, J. (eds) (1996) *Democracy and the Environment: Problems and Prospects*, Cheltenham: Edward Elgar.

Relation to the economy

Dresner, S. (2002) *The Principles of Sustainability*, London: Earthscan.

Ekins, P. (2000) *Economic Growth and Environmental Sustainability: The Prospects for Green Growth*, London: Routledge.

Pearce, D. (1995) *Blueprint 4: Capturing Global Environmental Value*, London: Earthscan.

Pearce, D. and Barbier, E.B. (2000) *Blueprint for a Sustainable Economy*, London: Earthscan.

Deep ecology

Naess, A. (1989) *Ecology, Community and Lifestyle: Outline of an Ecosophy*, trans. David Rothenberg, Cambridge: Cambridge University Press.

Part II

International engagement with sustainable development

3 Global governance and the United Nations environment Summits

Key issues

- Global environmental governance.
- UN environment Summits; Rio Earth Summit.
- International environmental regimes; Multilateral Environmental Agreements.
- Sustainable development indicators.
- The Millennium Declaration.
- Principles of good governance.
- Distinguishing structures and processes of governance.

The previous chapter discussed how the promotion of sustainable development is a global task: one that is directed towards the establishment of a more equitable relationship between the North and the South, the rich and the marginalized, the peoples of the present and the generations of the future. This chapter looks at how this understanding has been stimulated by and, in turn, has stimulated a new era of global environmental governance. It explores the factors that have contributed to the rise of global environmental governance, the key features of that system and their significance for the promotion of sustainable development.

The governance challenge

The Brundtland Report set an international political agenda for the promotion of sustainable development: the construction of effective, international cooperation to manage ecological and economic interdependences. This was a call both for new international institutions for global environmental governance and for changes in existing international agencies concerned with development, trade regulations and agriculture. 'The objective of sustainable development and the integrated nature of the global environment/development challenges pose problems for institutions, national and international, that were established on the basis of narrow preoccupations and compartmentalized concerns' (WCED 1987: 9). More specifically, Brundtland called for several changes in global environmental governance (Box 3.1).

Box 3.1 The governance challenge

- *Getting at the source*: supporting development that is economically and ecologically sustainable.
- *Dealing with the effects*: enforcing environmental protection measures and resource management.
- *Assessing global risks*: identifying, assessing and reporting on risks of irreversible damage to natural systems and threats to the survival, security and well-being of the world community.
- *Making informed choices*: supporting the involvement of an informed public, NGOs and the scientific community.
- *Providing legal means*: ensuring that national and international law keeps up with the accelerating pace and expanding scale of impacts on the ecological basis of development.
- *Investing in our future*: ensuring that multilateral financial institutions, including the World Bank, make a fundamental commitment to sustainable development and that bilateral aid agencies adopt a new priority and focus.

Source: adapted from WCED (1987: 20–1).

The Brundtland Report came at a time of growing concern about the inadequacy of national-level institutions and practices to address newly emerging global environmental problems. The Brundtland reconceptualization of the environmental problematic, from how best to manage pollution to how to promote sustainable development, contributed to the increased need for new arrangements for global environmental management. This is because promoting sustainable development is a quintessentially global project.

The development of global environmental governance

By the 1990s there were a substantial number of specific international environmental regimes, dealing with an array of environmental matters, from hazardous waste, ozone depletion and biodiversity loss to climate change. An international environmental regime exists when there are agreed-upon formal and informal institutional structures, principles, norms, rules and decision-making procedures and action programmes to address a specific environmental issue (Young 1997a). Their task is to secure negotiations, set standards of environmental management, especially of transboundary pollution, and find effective responses to the challenges presented by global environmental change. Examples of such regimes include those developed under the auspices of the United Nations to deal with climate change (the UNFCCC) and to address biodiversity loss (the CBD). Such conventions provide a general framework for action, while their associated technical protocols outline steps to address specific aspects of the problem. Regimes typically include multilateral environmental agreements (MEAs). Under the UNFCCC, the Kyoto Protocol allocates emission reduction targets to individual states in order to combat climate change. Similarly, under the CBD, the Cartegena Protocol on Biosafety deals with the safe transfer of living modified organisms.

While the internationalization of environmental management is primarily built upon negotiations and agreements between states, non-state actors, including environmental non-governmental organizations (NGOs) and economic actors, such as the Business Interest Association, play an increasingly significant role. The new structure of international environmental management reflects and, at the same time, helps to promote the rise of global civil society. This has changed the shape of international environmental politics, mobilizing both states and civil society at the international level.

This collective activity, including the development of international environmental regimes, their formal and informal institutional arrangements, their norms and principles, their MEAs as well as their conventions and protocols, when it is combined with the involvement of civil society, has resulted in the emergence of what has been termed 'global environmental governance'.

> Global environmental governance is the establishment and operation of a set of rules of conduct that define practice, assign roles and guide interaction so as to enable state and non-state actors to grapple with collective environmental problems within and across state boundaries.
>
> (Young 1997a)

The development of global environmental governance has led to the claim that the central role of the state in international environmental management is coming to an end. However, a more balanced view would see that, while international environmental politics is undoubtedly undergoing some new and rather innovative developments, this has not been at the expense of the state. The state still remains the primary actor in global environmental governance, even if it now plays that role in close collaboration with other actors. According to Young, these efforts both reflect and affect significant developments in the character of international society:

> Although states remain central players in natural resources and environmental issues, nonstate actors have made particularly striking advances both in the creation of environmental regimes and in efforts to make these regimes function effectively once they are in place. Environmental concerns are clearly one significant force behind the rising interest in the idea of global civil society.
>
> (Young 1997a: 2)

The United Nations environment Summits

The UN has played a particularly important role in the development of global environmental governance. There are now over thirty specialized UN agencies and programmes involved in the promotion of sustainable development at the global level. These include the Food and Agriculture Organization, the UN Development Programme (UNDP) and the UN Environment Programme (UNEP). Summits and conferences represent the public face of the UN's environmental engagement. They have resulted in numerous declarations, plans of action and sector-specific environmental conventions or laws. Summits have become central events shaping current international environmental policy and, more broadly, they are having a major influence on international relations, especially in relation to trade policy and North–South relations, and on the development of civil society at the global level.

The UN Conference on the Human Environment, held in Stockholm, Sweden, in 1972, marked the beginning of the new era of international cooperation on the environment. The conference was built upon two basic beliefs: poverty was a cause of environmental degradation and environmental problems could be solved by the application of scientific knowledge and technological know-how. The conference resulted in several significant developments. It helped to legitimize the view that environmental problems are of a global nature and it led to the formation of international structures and organizations to deal with them. One of the most important of these was the establishment of the UNEP in Nairobi, charged with the task of putting into practice the agreements reached in Stockholm. The conference also stimulated governments in over 100 countries to establish environmental ministries and agencies. It provided the impetus for the subsequent and rather rapid development of international environmental law, especially in relation to marine pollution, depletion of the ozone layer and transboundary movement of hazardous waste. It also marked the beginning of the explosive growth in the number of environmental NGOs, particularly those with an international remit.

Ten years later, the Stockholm + 10 Conference was held in Nairobi, Kenya. In contrast to the belief in the capacity of scientific and technological knowledge to solve environmental problems which marked the 1972 Stockholm Conference, the Nairobi Conference paid particular attention to the need to address the underlying economic and social causes of environmental problems. This led to the establishment of the World Commission on Environment and Development (WCED), chaired by Gro Harlem Brundtland. One of the concrete proposals made in the Brundtland Report was for the UN to hold an Earth Summit.

The Rio Earth Summit

On 22 December 1989 the General Assembly of the UN called for the convening of the UN Conference on Environment and Development, later to be known as the Earth Summit, held in Rio de Janeiro in 1992. The Summit was attended by the largest number ever of heads of state and of government, indicating the importance that had become attached by then to the subject of environmental deterioration and to the need to find ways of reconciling environmental protection with economic development policies at the international level.

The Rio Earth Summit focused on two key issues: first, the link between environment and development; second, the practical issues surrounding the promotion of sustainable development, especially the introduction of policies that balance environmental protection with social and economic concerns, particularly in the Third World. The Earth Summit resulted in an ambitious programme for promoting sustainable development (Box 3.2).

Box 3.2 Agreements reached at the Rio Earth Summit, 1992

- The Rio Declaration on Environment and Development.
- Agenda 21.
- The UN Framework Convention on Climate Change (UNFCCC).
- The UN Convention on Biological Diversity (CBD).
- The Forest Principles.

First, the Rio Declaration on Environment and Development was agreed, presenting 27 principles of sustainable development. These include the normative principles of common but differentiated responsibilities and the equity principles (inter-generational and intra-generational equity). The declaration also contains several principles of good governance, including the precautionary principle and the principle of subsidiarity, discussed on p. 71. Many of these principles address development concerns, but the declaration also dealt with principles concerning trade and the environment, and the role of civil society and social and economic groups in the promotion of sustainable development. The Rio Declaration also stresses the right to, and need for, development and poverty alleviation, especially in the Third World.

Second, Agenda 21 was agreed, which presented not only an astute analysis of the causes and symptoms of unsustainable forms of development, but an authoritative set of ideas on how to promote sustainable development in practice. Agenda 21 consists of forty chapters that outline action plans across a wide range of areas. It stresses the importance of bottom-up participation, especially community-based approaches, through Local Agenda 21 (LA21), discussed in Chapter 5 of this book.

Third, two legally binding conventions were signed at Rio, the UNFCCC and the CBD, both discussed in Chapter 4. The Convention to Combat Desertification also resulted from discussions held at Rio, but was not agreed until 1995. Fourth, the Forest Principles were also agreed, known formally as the 'Non-legally Binding Authoritative Statement of Principles for a Global Consensus on the Management, Conservation and Sustainable Development of all Types of Forests'. This outlined general principles of forest protection and sustainable management, while affirm-ing the sovereign right of the state to exploit forests. Some had hoped, however, for a legally binding forest convention, rather than the 'softer' Forest Principles. Agenda 21, the Rio Declaration and the Forest Principles are non-binding statements of intent, sometimes termed 'soft laws', which provide guidelines, or frameworks, for future development.

Fourth, the Rio Earth Summit led to the establishment of new institutions. Chief among these was the Commission on Sustainable Development (CSD), whose primary role is to monitor progress on the agreements reached at Rio. However, while it was initially envisaged as a high-level institution, it was linked with the Economic and Social Council rather than the General Assembly of the UN, a linkage that is seen as having limited its powers. The CSD has a very broad mandate and programme of work, which it pursues through annual sessions, less formal intersessional meetings, voluntary reporting from member states on implementation of Agenda 21 and through holding 'multi-stakeholder dialogues'. The CSD has also been charged with establishing sustainable development indicators, work also undertaken by the Organization for Economic Cooperation and Development (OECD), as discussed on pp. 57–9.

The CSD has been subject to considerable criticism, not least for being a slow and cumbersome organization. Several arguments can be made in its defence: from the outset, it was encumbered by a very broad mandate; it has to address a complex array of issues; it has to tread a politically delicate path between conflicting state and non-state interests; and, finally, very high expectations have been placed upon it by a myriad of groups and actors.

Despite these difficulties, the CSD has nevertheless been able to position itself at the heart of all the disparate, worldwide follow-ups to the Rio Summit (Bigg and Dodds 1997). By holding annual sessions to which governments are required to report, the CSD has, for example, stimulated the development of a system of national reporting, which, in turn, has helped to build a picture of global progress towards sustainable development. In this context, it can justifiably be claimed that, if the CSD did not exist:

> there would be a gulf between local and national implementation of Agenda 21 and global follow up. There would be much less possibility for the most significant bodies, including governments, international trade and financial institutions, etc. – to be held accountable to representatives of civil society.
>
> (Bigg and Dodds 1997: 34)

The usefulness of global reporting is dependent upon the development of an agreed upon set of indicators for measuring progress. Both the OECD and the EU have addressed this by devising sets of sustainable development indicators (Ekins 2000). Indicators transform the concept of sustainable development into something measurable. They measure three things: 'state' indicators measure the state of the environment; 'pressure' indicators measure the pressures or threats to which the environment is subject; and 'response' indicators measure societal responses to the problems (Box 3.3).

Box 3.3 OECD work on sustainable development indicators

Core environmental indicators

These track environmental progress and performance:

- Indicators of environmental pressures.
- Indicators of environmental conditions.
- Society's response.

Key environmental indicators

These track whether the public is informed and involved:

- Informing the general public.
- Providing key signals to policy makers.

---

continued

Promoting integration

These track whether environmental considerations are taken into account in policy making:

Sectoral environmental indicators

- Sectoral trends of environmental significance.
- Sectoral interaction with the environment.
- Economic and policy considerations.

Indicators derived from environmental accounting

- Environmental expenditure account.
- Physical natural resource accounts.
- Sustainable management of natural resources.
- Physical material flow accounts.
- Efficiency and productivity of material resource use.

Decoupling environmental indicators

These monitor progress towards sustainable development:

Macro-level decoupling indicators

- Decoupling environmental pressures from total economic activity, focus on climate change, air pollution, water quality, waste disposal, material and natural resource use.

Sector-specific decoupling indicators

- Production and use of resources in specific sectors.

Source: adapted from OECD (2003).

Environmental indicators have to be distinguished from sustainable development indicators. Sustainable development indicators address social, economic, ethical as well as environmental considerations. They take greater account of the synergistic effects of behaviour, especially the ways in which ethical considerations influence behaviour across the social, economic and environmental spheres (Box 3.4).

Indicators provide ways of monitoring progress and of comparing that progress internationally, or across sectors or over time. Nevertheless, while indicators appear to present a somewhat technical approach, it is important not to see them

---

Box 3.4 A framework for sustainable development indicators

- *For the environment*: can its contribution to human welfare and can the human economy be sustained?
- *For the economy*: can today's level of wealth creation be sustained?
- *For society*: can social cohesion and important social institutions be sustained?
- *Ethically*: do people alive today value other people and other life forms, now and in the future, sufficiently highly?

Source: Ekins (2000: 104).

---

as 'neutral' management tools. The use of indicators, for example, to set targets for policy, is inevitably a political exercise that involves difficult, and rarely consensual, political choices (Redclift and Woodgate 1997). Therefore, difficult decisions have to be made about what environmental functions or quality should be maintained, and at what level (Ekins 2000). A good example is provided by climate change. For the climate to be stabilized, atmospheric concentrations of greenhouse gases have to be kept below a certain level. This requires understanding the critical load of the ecosystem, setting indicators of planetary health and reaching agreement on maximum levels of emissions of these gases. All these tasks require value judgements, hard political negotiations, and agreement on priorities. Faced with the prospect that restrictions on emissions may place undue burdens on industry, a government may decide that other public policy objectives, such as employment, are more important than environmental sustainability and it may, therefore, wish to set relatively high emission levels.

Conflicting views on the Rio Earth Summit

Despite the many agreements reached at Rio, and the range of outputs, conventions, laws, organizations and activities that it has spawned, the Rio Earth Summit is not without its critics. The two Bush administrations in the US, for example, have repeatedly questioned the sustainable development principles that form the basis of the Rio Declaration, not least the upholding of the precautionary principle as a guide to international environmental law. Far from remaining an abstract debate, the US administration's rejection of the precautionary principle has proved highly influential, shaping its position on several key international environmental policy issues, including global climate change, as discussed in Chapter 4. The US administration has also held the Rio process to ransom, as it were, by consistently refusing to meet the financial obligations laid down at Rio, issues discussed further in Chapter 7.

Criticism has not only originated from within the US administration. On the contrary, some of the most scathing criticism of both the Rio Summit and the entire international governance process that it has spawned has come from within the radical Green movement. *The Ecologist* magazine has argued that the very premise of the Earth Summit was flawed (*Ecologist* 1993). By focusing attention on *how* the environment should be managed, it basically addressed the wrong question. The real question, they argue, was not how but *who* will manage the environment and in *whose* interest. They raised the issue of how the 'global' environment was defined at Rio: designating certain issues as global, such as climate change and biodiversity loss, and others as local, such as desertification, was seen to reflect the interests of the politically and economically powerful nations of the industrialized world. *The Ecologist* asked: whose common good is being protected by the Rio process? In reply they argue that powerful states use institutions such as the UN to transform their own state interests into international agreed-upon, environmental norms and governance systems (Box 3.5). Particularly at the negotiating stage of regime formation, states have a powerful interest in ensuring that considerations of costs, benefits or problems of domestic implementation remain dominant factors in shaping outcomes (Breitmeier 1997). Radical Greens argue that this is precisely what was happening at Rio and has continued since.

Box 3.5 *The Ecologist*'s radical repudiation of the Rio Earth Summit

- It is not poverty which is the root cause of environmental degradation, but the Western style of wealth.
- Overpopulation is caused, not cured, by modernization, as it destroys the traditional balance between people and their environment.
- The 'open international economic system' of the Rio Declaration will extinguish cultural and ecological diversity.
- The problem of pollution is to be solved not by pricing the environment (market solutions) but by reversing the enclosure of the commons, that is, the process whereby the common resources upon which people have traditionally depended are being brought under the remit of commercial interests, or commodified.
- The call for more 'global management' constitutes another example of Western cultural imperialism.
- The idea that developing countries urgently need the transfer of Western technology smacks of arrogance. It assumes that ignorance is characteristic of Third World people.

Source: adapted from Pepper (1996).

In addition to the radical Green critique of Rio and its related developments, one that draws heavily upon a Third World perspective, there is another, more fundamental, critique put forward by radical environmentalists and Green theorists. Because this critique takes issue not just with the Rio Earth Summits and their related developments but the whole development of global environmental governance for the management of the environment, it is discussed in a separate section at the end of this chapter.

The UNGASS New York Summit: Earth Summit + 5

In 1997 fifty-three heads of state or government and sixty-five ministers of the environment and other areas attended a United Nations General Assembly Special Session (UNGASS) to review the implementation of Agenda 21. This event is known as the Earth Summit II, Earth Summit + 5 or simply as UNGASS and was held in New York in June 1997 (Box 3.6).

Box 3.6 The Earth Summit + 5 objectives

- To revitalize and energize commitments to sustainable development.
- To take stock of progress since Rio.
- To define priorities for the post-1997 period.
- To raise issues addressed insufficiently by Rio.

Source: Osborn and Bigg (1998).

Reports on the state of the world's environment prepared for UNGASS painted a dismal picture. They showed that the global environment had continued to deteriorate at an alarming rate, with rising levels of greenhouse gas emissions, toxic pollution and solid waste. Renewable resources, notably fresh water, forests, topsoil and marine fish stocks, continued to be used at rates that are unsustainable. World leaders began the UNGASS meeting with the sobering realization that, five years after the Rio Earth Summit, the planet's health was worse than ever.

Yet, despite this knowledge, there were no major breakthroughs at New York. Discussions became bogged down in North–South differences on how to finance sustainable development globally. Pledges made at Rio by donor countries in the North to increase official development assistance (ODA) and make environment-friendly technologies available to developing countries had not been kept. Rather,

ODA had declined from an average 0.34 per cent of donor country gross national product in 1991 to 0.27 per cent in 1995. Many in the South saw this as a breakdown of the global partnership for development declared at the Rio Earth Summit.

The hotly contested debate on whether there should be a legally binding forest convention also undermined the Earth Summit + 5. Canada and the European Union strongly favoured a new treaty, but the United States, Brazil, India and most major environmental organizations were opposed. Many Third World countries did not want to see a legally binding forest convention, fearing that it would hamper their development plans. In contrast, major environmental organizations opposed a convention, fearing that it would at best offer only weak measures and, at worst, bring forests within the remit of global environmental management regimes that would, paradoxically, open up forests to commercial logging development under the guise of introducing sustainable practices. This is an example of the closure of the commons discussed in the *Ecologist* critique.

The final document adopted by delegates disappointed many. It contained few new concrete commitments on action. Ambassador Razali Ismail of Malaysia, President of the General Assembly and Chair of the Special Session, in his address to UNGASS, on 23 June 1997, expressed his disappointment as follows:

> We are familiar with the tactics being played. Posturing, spinning declarations of intent, pointing the finger at others, pandering to interest groups, weighing short-term profits and immediate electoral gains, and emphasizing the need for clearer definitions, dialogue or information-gathering. These prevent plans of action from truly being operationalized into programmes of implementation.
>
> (Ismail, quoted in Third World Network 2004)

However, the UNGASS meeting was significant in that NGOs were able, for the first time, to deliver speeches to the plenary sessions and to have access to ministerial-level consultations. This is evidence that the system of environmental governance promoted by the UN encourages the participation of private and voluntary groups (Elliott 2002). There are several reasons for this encouragement. First, it lends democratic legitimacy to UN proceedings. Second, it gives the UN access to vital information, as many NGOs have built up considerable expertise and local knowledge on a wide range of environmental issues. Finally, it helps form new allies with which to construct implementation partnerships for sustainable development, as seen in the development of Type II partnerships at the WSSD, discussed below.

Box 3.7 Outcome of Earth Summit + 5

- Confirmation of the political commitment to the promotion of sustainable development.
- Frank assessment of progress.
- Recognition of the key role of the UN in this area.
- Confirmation of UNCED targets and commitments for ODA.
- Clarification of the role of specific institutions.
- More focused programme of work of the CSD.
- Adoption of a Programme for the Further Implementation of Agenda 21.
- Continuation of work on forests and consideration of a possible legally binding instrument in this area in the future.
- Beginning of intergovernmental process within CSD on fresh water and energy.
- Better understanding of issues around sustainable development and tourism, transport, information and changing production and consumption patterns.
- A number of new practical agreements in special areas such as a worldwide phase-out of lead from gasoline.

Source: adapted from Osborn and Bigg (1998).

A parallel Rio + 5 Forum was held at the same time as the Earth Summit + 5, and organized by the Earth Council. The Earth Council is an environmental NGO set up by Maurice Strong, who chaired the negotiations at the 1992 Rio Summit. This meeting pointed to the many areas of rapid and enhanced environment degeneration, to the rise in poverty and inequality and to the many key environmental resources that are becoming scarcer, especially water. It highlighted the need for industrial countries to instigate policies aimed at promoting new patterns of sustainable consumption and production.

The alternative forum struck a chord with the ongoing debate about the moral, political and financial responsibilities of the industrial world to address their 'ecological debt' – that is, the negative ecological legacies that it has bequeathed both to the South and to future generations. The Brundtland Report and the Rio Declaration had made it abundantly clear that attempts to promote sustainable development require the North to take these responsibilities seriously and act upon them. This belief is also reflected in the UN Millennium Declaration, a declaration that brought together the agreements reached at the numerous UN world conferences that have been held over the previous ten years (Box 3.8). However, declaratory political statements are but one step. Funding, as well as changes in the way in which the international economy, including international trade, is organized is also an essential prerequisite for our sustainable future. Some would argue that these call forth fundamental structural changes in the international economic and political system.

Box 3.8 The Millennium Declaration

- The Millennium Declaration of the UN, agreed at the Millennium Summit in 2000, summarized the agreements and resolutions of the UN world conferences held during the previous ten years to establish the Millennium Development Goals. These are seen as benchmarks for measuring actual development.
- There are eight Millennium Development Goals and the environment is an essential component of them. The first seven are about poverty reduction and improving health. These goals are directly linked with the promotion of sustainable development.
- Goal No. 7 is particularly important for the promotion of sustainable development. It has several targets: mainstreaming the environment into policies and programmes (environmental policy integration), reversing the loss of environmental resources and improving access to environmental services.
- It also aims to halve the proportion of people without sustainable access to safe drinking water by 2015. It also seeks to achieve a significant improvement in the lives of at least 100 million slum dwellers by 2020.
- Goal No. 8 stresses that the achievement of these goals requires a global partnership for development. The Millennium Development Goals are reflected in the Johannesburg Plan of Implementation.

Source: adapted from http://www.un.org/millenniumgoals/, accessed 27 April 2005.

World Summit on Sustainable Development 2002: Earth Summit + 10 or Rio − 10?

The third Summit, the World Summit on Sustainable Development (WSSD), was held in Johannesburg in 2002. The WSSD had two goals: to hold a ten-year review of the 1992 Earth Summit and to reinvigorate the global commitment to sustainable development. The WSSD was the biggest event of its kind organized by the UN to date, both in terms of its scope and in terms of its complexity. There was also a parallel conference, the Global People's Forum, organized by and for civil society.

Several inputs fed into the WSSD: the outputs of the UN environmental conferences held up to that date; the UNCED documents and reports; and, more specifically, the Millennium Declaration, the WTO Doha Declaration and the Monterrey Consensus (see Chapter 7). Like the Rio conferences, the WSSD was preceded by four preparatory meetings, known as 'PreComs'. Twenty-two reports on the implementation of Agenda 21 were presented at the first PreCom. They highlighted the limited achievements in global efforts to promote sustainable

Figure 3.1 *The UN World Summit on Sustainable Development, Johannesburg 2002, faced wide-ranging and incompatible expectations*

Courtesy United Nations

development in the years since 1992. They showed a fragmented approach, lack of progress in addressing unsustainable patterns of production and consumption, inadequate attention to the core issues of water and sanitation, energy, health, agriculture, biodiversity protection and ecosystem management (issues known under the acronym WEHAB); lack of coherence between policies on finance, trade, investment, technology and sustainable development; insufficient financial resources; and absence of mechanisms for technology transfer. These reports helped focus the attention of the Summit on the practical issues surrounding implementation of strategies. However, many argued that the Summit was doomed to failure even before it started, because national representatives and civil society organizations came to Johannesburg with very different and sometimes incompatible agendas. The result was a daunting, long 'wish list', which Johannesburg could only fail to meet (Box 3.9).

Box 3.9 Agreements reached at the WSSD, 2002

- Johannesburg Declaration on Sustainable Development.
- Johannesburg Plan of Implementation.
- WEHAB initiatives.
- Type II partnerships.

The Summit resulted in the Johannesburg Declaration on Sustainable Development. This refers to the need to promote sustainable development through multi-level policy actions, adopting a long-term perspective and encouraging broad participation. However, the declaration is more in the nature of a political compromise, desperately needed to lend weight and legitimacy to the closing moments of the Summit. The declaration lacks the intellectual sophistication and authority that the Rio Declaration still commands (Hens and Nath 2003). Unlike the Rio Declaration, the Johannesburg Declaration is also seen as unlikely to lead to new international negotiations or legal conventions.

The second output, the Joint Plan of Implementation, is the core document of the WSSD, and describes how already existing commitments and targets might be met. One of these is the commitment to achieve sustainable harvesting practices in the world's fisheries by 2015. The plan prioritizes WEHAB initiatives. Agreements in relation to renewable energy use and addressing biodiversity loss were also reached but are regarded as particularly weak, especially since no new commitments were made to increase aid and deal with debt. More generally, the plan is seen to lack innovative thinking, in particular the section that deals with climate change. So too are discussions on emerging issues, especially globalization and trade, where the plan calls only for 'an examination' of the relationship between trade, the environment and development. It failed to invoke the precautionary principle in dealing with potential problems at the interface between the environment, trade and development, nor does it present any new insights or clarity into the mechanisms linking trade, the environment and sustainable development (Hens and Nath 2003). In the words of one critic:

> This Summit has failed the poor and vulnerable peoples of the world. It has not reached agreement on the radical action – with clear timetables and targets – needed to tackle the world's environmental problems, from climate change and renewable energy to forest and species loss. The world's Governments must agree to meet again and determine that next time they will do better.
>
> (Ricardo Navarro, Chair of FoE International, in FoE 2002)

The most innovative, but potentially most problematic, implementation strategy agreed in the plan is the so-called Type II partnerships. While Type I actions are negotiated by governments, Type II partnerships are the result of groups of countries, their national or sub-national governments, the private sector, especially the business community, and civil society actors, backed by financial commitments, working in partnership to agree international, voluntary agreements for specific, concrete initiatives. Several multi-stakeholder projects were announced at the WSSD, including Capacity 2015 of the UNDP, the Water and Energy Partnership of the EU, the UNESCO Encyclopaedia of Life Support Systems, and the International Youth Dialogue on Sustainable Development of the Global Youth Network.

Partnership arrangements recognize the importance of including a wide range of economic and social actors in the promotion of sustainable development. They help secure participation through their focus on *concrete* development initiatives. However, critics have pointed to a danger that partnership arrangements, especially those involving the private, business sector, may result in the commodification of the environment, turning more and more environmental assets into resources that are to be managed, albeit more responsibly. There is also concern that state authorities may lose authority over their sustainable development policies and programmes, particularly in selected areas such as forests, by allowing control to shift to powerful business or industry interests, and groups that do not prioritize the promotion of sustainable development. Political corruption and the lack of democratic transparency and accountability can also enhance these dangers. The conflict over Type II partnerships can be seen in the different opinions expressed over one such initiative, the Congo Basin Forest Partnership (Box 3.10).

Box 3.10 The Congo Basin Forest Partnership

The Congo basin contains a quarter of the world's tropical forest. It is a region of extraordinary biological richness, but the Congo basin forest is being degraded at the rate of 2 million acres every year.

The partnership is an initiative by the United States, the six governments of the Congo basin, other partner governments, conservation and business groups, and organizations representing civil society. It was launched by the US Secretary of State, Colin Powell, at the WSSD in Johannesburg 2002. Action is focused on eleven key landscapes in Cameroon, the Central African Republic, the Democratic Republic of Congo, Equatorial Guinea, Gabon and the Republic of the Congo.

Aim

The aim of the Congo Basin Forest Partnership is to manage the Congo basin forest in a sustainable fashion. The partnership provides support for:

- a network of national parks and protected areas;
- the issuing of forestry concessions to logging companies;
- the creation of economic opportunities for communities that depend upon the conservation of the outstanding forest and wildlife resources of the Congo basin.

---

continued

The US government will invest in the partnership through a $12 million per year increase within USAID's Central African Regional Program for the Environment. From 2002 to 2005, the US plans to invest up to $53 million in the partnership.

Partners

Governments: United States, Cameroon, Central African Republic, Democratic Republic of Congo, Equatorial Guinea, Gabon, Republic of Congo, United Kingdom, Japan, Germany, France, Canada, South Africa and the European Union.

International Organizations: World Bank, International Tropical Timber Organization and World Conservation Union.

Civil Society: Jane Goodall Institute, Conservation International, Wildlife Conservation Society, World Wildlife Fund, World Resources Institute, Forest Trends, Society of American Foresters, American Forest and Paper Association, Association Technique Internationale des Bois Tropicaux and the Center for International Forestry Research.

Conflicting interpretations

The US contributions to the Congo Basin Forest Partnership will promote economic development, alleviate poverty, and improve local governance, through natural resource conservation programs.

(Bureau of Oceans and International Environmental and Scientific Affairs)

The Congo Basin Initiative is a key example of the Bush Administration's flawed partnership approach. While supposedly benefiting forest protection and management in the highly biodiverse Congo Basin, the initiative will actually put more money into flawed programmes that have not reduced illegal logging, empowered local communities or enabled sustainable forest management. The US has also dismissed concerns of local environmental groups about corruption in these countries and the close collusion between government officials and timber barons.

(Friends of the Earth)

Sources: adapted from http://www.cbfp.org/en/about.aspx, accessed 6 January 2004; Bureau of Oceans and International Environmental and Scientific Affairs (2002); FoE UK (2002).

Conflicting views on Type II partnerships reflect more widespread disagreements about the role played by the US in the UN environmental Summits. Colin Powell was heckled at the WSSD, where the US tried to block agreement on substantive timetables and targets on several key issues. Ever since Rio in 1992, the US has rejected targets and timetables, an action that has considerably weakened many of the agreements reached at the UN Summits. Friends of the Earth have argued that the US position is particularly egregious, given the disproportionate share of global resources it consumes and the environmental damage it does (FoE UK 2002). In addition, there were disputes with the US about the relationship between MEAs and free trade agreements. The US pushed hard for the WTO agreements to take precedence over environmental agreements, against the wishes of the EU and the G-77. This points to the marked tension that exists between the international trade and environmental regimes, an issue that is discussed further in Chapter 7.

However, another, more positive interpretation of the WSSD has also been given. This stresses the new level of dialogue in Johannesburg between all the stakeholders, especially between governments, civil society and the private sector, as reflected in the Type II partnerships:

> [T]he Summit was anything but a complete failure. [I]t was an opportunity . . . to exchange ideas and information on achieving sustainable development, and to strengthen networks.
>
> Ken Ruffing (Head of OECD Environment Directive) 2003

> The Summit will be remembered not for the treaties, the commitments, or the declarations it produced, but for the first stirring of a new way of governing the global commons – the beginning of a shift from the stiff formal waltz of traditional diplomacy to the jazzier dance of improvisational solution-orientated partnerships that may include non-governmental organisations, willing governments and other stakeholders.
>
> Jonathan Lash, President of World Resource Institute
> (quoted in UNDESA 2002)

This interpretation is closely linked with the view that the WSSD was, more than anything else, an act of international environmental diplomacy.

Box 3.11 Summary of main events in global environmental governance

UN environment Summits

- UN Conference on the Human Environment, Stockholm, 1972.
- UN Stockholm + 10, Nairobi, 1982.
- UN Conference on Environment and Development, Rio de Janeiro, 1992 (Earth Summit).

---

continued

- UN General Assembly Special Session to Review Agenda 21, New York, 1997.
- World Summit on Sustainable Development, Johannesburg, 2002.

Some related conferences

- International Conference on Population and Development, Cairo, 1994.
- World Summit for Social Development, Copenhagen, 1995;
  WSSD + 5, 2000.
- Fourth World Conference on Women, Beijing, 1995;
  Beijing + 5, New York, 2000.
- UN Conference on Human Settlements, Habitat I, Vancouver, 1978;
  Habitat II, Istanbul, 1996;
  Habitat III, Istanbul + 5, 2001.
- First Global Ministerial Environmental Forum, Malmö, Sweden, 1999.
- UN Conference on Financing for Development, Monterrey, 2002.

Major reports and declarations

- First report of the Club of Rome, 1972.
- UNEP World Conservation Strategy, 1980, first comprehensive policy statement linking conservation and sustainable development.
- WCED, *Our Common Future*, 1987.
- Rio Declaration on the Environment and Development, 1992.
- Agenda 21, 1992.
- 'We the Peoples', millennium report of the UN Secretary General, 2000.
- WTO Doha Declaration, 2001.
- Monterrey Consensus, 2002.
- Johannesburg Declaration on Sustainable Development, 2002.

Development of international environmental law

- Vienna Convention on the Protection of the Ozone Layer, 1985.
- Montreal Protocol on Substances that Deplete the Ozone Layer, 1987.
- UN Framework Convention on Climate Change, 1992;
  Kyoto Protocol, 1997.
- UN Convention on Biological Diversity, 1992;
  Cartagena Protocol on Biosafety, 2000.
- Forest Principles, 1992.
- UN Convention to Combat Desertification, 1995.

Related initiatives

- Rio + 5, New York, 1997.
- Earth Charter, 2000.

Principles of good governance

At the start of this chapter the Brundtland call for a new system of global environmental governance was mentioned. The Brundtland Report was aware of the need to address the imbalances in the current structures of global governance and to create a more inclusive governance system. For Brundtland, an important guiding principle for such reform was the fair and equitable distribution of bargaining power, so as to ensure that the voice of the world's poor is heard and indeed reflected in international decisions and outcomes.

This call has led to the elaboration of a set of principles for good governance practice. Some of these were elaborated at the Rio Earth Summit, where the Rio Principles outline several of the core elements of good governance for sustainable development. The WSSD Plan of Implementation also dealt with the issue of good governance (Box 3.12).

Box 3.12 UN principles of good governance

General principles of good governance

- The rule of law.
- Transparency and accountability.
- Effectiveness and efficiency.
- Subsidiarity: actions should be taken at the appropriate level of government.
- Participation and responsiveness to the needs of stakeholders.
- Gender equity.

Specific principles of good governance

- Precautionary principle.
- Principle of common but differentiated responsibilities.
- Ecosystems approach.
- The 'polluter pays' principle.
- Principle of environmental policy integration.

Several of these principles of good governance are discussed in detail in this book. The precautionary principle is discussed in Chapter 4, while EU efforts to promote environmental policy integration are discussed in Chapter 6. Chapter 2 explored how the principle of common but differential responsibilities has shaped international environmental law, and it looked in some detail at the principles of participation and of gender equality. The governance principles have generated severe controversy among the states participating in the UN Summits. At the

WSSD the conflict was so extreme that a conflict resolution group had to be set up. As a result, references to adopting the precautionary principle with regard to biodiversity conservation were dropped, the US and Japan having worked hard for the words 'the precautionary approach'. Similarly, with respect to the Principle of Common but Differentiated Responsibilities, the industrial countries argued that it should refer only to environmental, not development, matters, creating an impasse between industrialized countries and the G-77 governments.

The UN Summits have also led to a set of expectations about the conduct of global environmental governance regimes, or about what ought to be the case. Indeed, much of the discussion of good governance is of a highly normative nature. The principles of good governance identified by the UN-accredited NGO the World Humanity Action Trust provide a good example (Box 3.13). The UN system for the promotion of sustainable development put in place since the Brundtland Report is represented in Figure 3.2.

Box 3.13 World Humanity Action Trust principles of good governance

- Enable science and technology to inform policy making and policy implementation at local, national and international levels.
- Increase funding and programmes for capacity building in policy making and implementation.
- Promote vision, values and, above all, joined-up thinking (horizontal, vertical and temporal) to secure sustainability.
- Ensure the creation of inclusive organizations, that are willing to delegate, i.e. accept the principle of subsidiarity, that are resourced at realistic levels (from sources preferably independent of national treasuries), and which command the respect of individuals, civil society, technical/managerial expertise and nation-state politicians.
- Encourage clustering across the social/cultural spectrum, across environmental, economic and political divisions and avoid the environmental ghetto.
- Take an ecosystem approach and a problem-orientated one.
- Reduce territoriality in favour of collective thinking and action.
- Demand greater public accountability on a global scale.

Source: World Humanity Action Trust (2001).

Significance of UN Summits

Sceptics may argue that UN conferences, such as the Rio and Johannesburg Summits, were merely temporary media events and do not have significant effects.

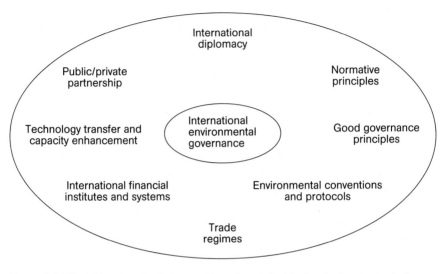

Figure 3.2 *The UN system for the promotion of sustainable development penetrates into all areas of international governance*

However, issue can be taken with this interpretation, arguing that the environmental Summits and conferences organized under the auspices of the UN have served several important functions. To see the range and significance of these functions, it is useful to draw upon the work of Haas (2002), who has argued that the functions of conference diplomacy include:

- agenda setting;
- popularizing issues and raising consciousness;
- generating new information and new challenges for government;
- providing general alerts and warning of new threats;
- galvanizing administrative reform;
- adopting new norms and doctrinal consensus;
- promoting mass involvement.

There is little doubt that the UN environment Summits and conferences have contributed to each and every one of these developments. They have established the agenda of global environmental politics around the aim of promoting sustainable development. This has shifted attention from earlier preoccupations with the environment as merely a technical issue of pollution control to the current efforts to reconcile social, economic and ecological goals. As a result of the UN's engagement, promoting sustainable development has now become a norm of global environmental politics.

Closely associated with this has been the articulation of a common set of normative principles as well as good governance principles. The system of global environmental governance that has developed since Brundtland is marked by its adherence to an increasingly sophisticated web of normative and governance principles.

UN Summits and conferences have resulted in a pattern of increasing negotiations, cooperation, participation and institutional development at the international level. The Rio Conference was particularly important in this respect. It also helped in the formation of new environmental ministries and agencies in several countries across the globe. This creates order in the management of the environment at the international level.

Conferences and Summits have also helped develop 'soft law' norms, which become, over time, acceptable to a large group of nations and which subsequently evolve into 'hard law'. The UN Summits have helped to build up the body of MEAs, including environmental conventions, such as the UNFCCC, which resulted in the Kyoto Protocol and the UNCBD, and its subsequent Cartagena Protocol on Biosafety.

The reports prepared for the PreCom meetings, in advance of Rio, UNGASS and the WSSD, have helped identify areas for priority action. Over time, it is hoped that similar attention can be paid to implementation, as the conferences of the parties (CoPs) reach agreement on rules, obligations, reporting requirements, and resource transfers to aid implementation, particularly in the Third World.

Global environmental diplomacy has been, at least in part, democratized as a result of Summit practice. This has occurred through the work of Agenda 21, which requires governments to involve major societal groups in decision making and implementation strategies in pursuit of sustainable development and to give access to information on environmental problems and policies. In addition, the UN has officially accredited thousands of transnational NGOs and given them a voice at Summit meetings.

Despite the positive functions of the UN Summits, conferences and follow-up activities, these events are, nevertheless, subject to very valid criticisms. These include concern about the exclusion of certain issues from centre stage in the agenda of global environmental policy and about the ability of powerful states to transform their interests into internationally agreed-upon governance systems and regimes. The ability of the predominant global power, the US, to hold the UN to ransom is a case in point, by refusing pay to dues and, when all else fails, opting out of its conventions.

Much of the criticism has been directed at the inadequacies of the UN environmental governance structures. Concern has been expressed that the agenda of international environmental politics that it fosters is not sufficiently inclusive, that legislation is not sufficiently strict and that funding has not been forthcoming, in particular from the industrialized world. Accepting the validity of the system, the desire is to make that system both more efficient and more effective. However, there is also another, more fundamental critique, one that stems from within the radical edge of the environmental movement and from within radical Green theory, one that rejects the system of global environmental governance entirely.

Radical environmentalism and the rejection of the system of global environmental governance

The UNCED process has helped consolidate two distinctive ways of dealing with the relationship between the environment and development:

- through the mobilization of intergovernmental organizations and, increasingly, through partnerships from within civil society;
- through the creation of international agreements, including hard laws.

For radical Greens this has resulted in the development of an international environmental governance system that is increasingly preoccupied with issues of environmental resource management: how to use nature and, at the same time, limit and preferably reverse environmental degradation and the exhaustion of resources (Redclift 1999). Drawing upon sociological analysis, this development is seen as part of a general trend in modern society, or modernity, towards the increased *rationalization* of everyday life. Modernity refers to the sum of a series of complex historical processes operating at four interrelated levels. At the political level, it is characterized by the rise of the secular state and polity; at the economic level, it sees the rise of the global capitalist economy; within the social level, there is the formation of classes and an advanced sexual and social division of labour; finally, the cultural level sees a transformation from a religious to a secular culture (Hall *et al.* 1992). The rationalization of everyday life is a process whereby science, technology and also the policy process become dominated by concerns about means and procedures. This has led to an understanding of the environmental problematic as a technical problem whose solution involves the application of the correct techniques, within the context of bureaucratic institutions. Thus, to address the world's environmental problems, new institutions for international environmental governance are needed, which, in turn, require negotiated international environmental agreements and technical tools to measure progress and monitor compliance. This means that the environmental agenda becomes crowded by debates about the means of achieving given ends, rather than debates about the ends themselves. As a result, normative issues become subordinated to concerns about techniques.

A managerialist approach does not call for fundamental changes in environmental and other values or existing patterns of production or consumption. Rather, it shifts environmentalism from a critique of lifestyles and consumption patterns to a question of devising better managerial strategies and more effective and efficient institutional control over the environment (Sachs 1999). For radical Greens, this approach tames the agenda of environmental politics and it has led some to reject the whole notion of sustainable development as flawed managerialism.

Similarly, a political science perspective has been used to criticize the emergence of the sustainable development agenda in international politics. First, while acknowledging that much of the discussion in the Brundtland Report was framed within an implicit social democratic political framework, based on a vague 'communitarian' ethos, the sustainable development agenda of international institutions is criticized because it has shifted focus. Here there is less interest in sustaining a 'common future' than concern to push particular interests to the forefront of the international stage (Barnes 1995). This means that little is done to develop an approach that deals with the environmental problematic as an issue of common purpose (Dryzek 1990). Second, the social democracy of Brundtland has often been replaced by a neoclassical, free-market perspective, as found, for example, in the highly influential environmental economics of Pearce. In Pearce's work (1994, 1995; Pearce and Barbier 2000) the framework of market liberalism is taken for granted, which places the individual, not the community, at the centre of analysis. This approach has strongly influenced the weak sustainable development agenda adopted in both the UK and Australia (Barnes 1995). The controversial sustainable development strategy of the UK government, *A Better Quality of Life* (1999), for example, placed emphasis on maintaining high levels of economic growth while paying less attention to objectives about the environment, society and the use of resources, and even less to the UK's approach to LA21. In this political science view, the sustainable development agenda has become, in the hands of governments and international agents, part of a legitimizing project for the neo-liberal approach towards politics and economics. It is neo-liberal in that it promotes and justifies the use of markets to manage the environment and strengthens an individualistic approach to what is in effect a collective problem.

However, it is possible to embrace the agenda of sustainable development, while remaining critical of its expression in and through international governance regimes. To develop that argument further, a distinction can be made between the *institutions of governance* and *governance processes*. Much of the discussion in this chapter has focused on exploring the institutions of governance. It has looked at how governance institutions have developed through the holding of regular meetings, the formation of new administrative procedures and specific institutions, the negotiation of new laws, and the establishment of monitoring and reporting systems. These institutional arrangements were shown to have become, over time, more open, more democratic and more principled in their approach. However, a different picture emerges when governance processes are examined – that is, when attention is paid to the context in which these institutional arrangements operate. Power relations have been shown to still play a major role in shaping the output of the UN Summits. The uneven distribution of both economic and political power, whether in relation to the industrialized world (the US in particular) and the Third World or in relation to economic interests and social groups, remains manifest when attention is focused on governance processes. This perspective shows that governance of the environment is still mediated by powerful and inequitable international and national processes, especially manifest in trade and finance. For those who wish to unfold the radical potential of the sustainable development agenda, action is needed to deal with the uneven distribution of economic and political power. The importance of this action will be seen when the factors that limit the capacity of local actors to effectively engage in LA21 (Chapter 5) and the ability of developing countries to promote sustainable development (Chapter 7) are explored.

Conclusion

The rise of global environmental governance, of which the UN Summits form a key part, is the subject of considerable debate (Box 3.14). In short, the UN environment Summits can be interpreted in two conflicting ways. First, they can be seen as acts of global diplomacy, in which case they have served several important functions. Alternatively, they can be judged from a radical environmental stance as ways in which capitalism can escape from its environmental crisis through the technical management of the environment and the use of markets. In this latter perspective, they are to be viewed with deep scepticism and mistrust.

At the heart of the discussion about global governance is the distinction between governance structures and governance processes. While governance structures have been opened up by the UN engagement with sustainable development, especially at the international level, these structures continue to operate within governance processes. Here the differential exercise of political and economic power, be it at the international or the local level, limits the ability of actors to engage effectively in the promotion of sustainable development. Recognition of this difference lies at the heart of the Brundtland call for good governance practice. An important guiding principle in global governance reform is the fair and equitable distribution of bargaining power, so as to ensure that the influence and voice of the world's poor are heard and indeed reflected in international decisions and outcomes that seek to promote sustainable development.

Box 3.14 Summarizing the role of the UN Summits in international environmental governance

Positive aspects	*Negative assessment*
Facilitate international negotiations	Lack substantive output
Help increase cooperation	Promote the rhetoric of cooperation
Encourage participation	Allow only limited or elite participation
Strengthen global focus	Particularistic in focus and ignore the local
Develop mutual understanding	Lowest common dominator agreements; national interests remain dominant
Lead to treaties and binding conventions	Overly cautious, inadequate targets
Establish institutional or global governance	Ineffective, lack competence and political clout, poorly funded

---

Summary points

- The UN has played a key role in shaping the international response to the environmental crisis and in structuring that response around the norm of sustainable development.
- This has facilitated the development of an international governance structure aimed at more effective environmental management. As a result, environmental governance is no longer coterminous with a delimited political territory, nor is it seen as the exclusive business of governments.
- While states remain key actors in the UN system, the UN has helped to open up international environmental governance to a wide range of groups from within civil society and the economy. This has changed the face of international environmental politics.
- The role of the UN is subject to two conflicting interpretations. On the one hand, it is seen as making a positive contribution to our collective future, by structuring the legal, institutional and political engagement with sustainable development at the international level.
- In contrast, the UN can be criticized for being a management agent, helping to promote a system of global environmental governance preoccupied with

means and not ends. This displaces a more fundamental critique of the flaws of development policies, the fallacy of environmental management and the role of the market in promoting individual preferences, to the detriment of the common good.

• At the root of these conflicting interpretations lies deep conflict over whether sustainable development is a tool for the construction of a radically new environmental future or whether it is to be rejected out of hand as little more than an anthropocentric management tool, useful to help capitalism – or at least the neo-liberal expression of capitalism – to find a way out of its environmental crisis.

• It is helpful to return to the Brundtland formulation and to make a distinction between the formation of new institutions of governance and the processes of governance, which remain constrained by the wider system of international political and economic power. Embedded in the Brundtland plea for good governance is the recognition that promoting sustainable development requires changes at this more fundamental level.

Further reading

UN understanding of sustainable development

World Commission on Environment and Development (1987) *Our Common Future*, Oxford: Oxford University Press.

Earth Summits and their agreements

Grubb, M., Koch, M., Munson, A., Sullivan, F. and Thomson, K. (1993) *The Earth Summit Agreements: A Guide and Assessment*, London: Earthscan.

Hens, L. and Nath, B. (2003) 'The Johannesburg Conference', *Environment, Development and Sustainability*, 5: 7–39.

Osborn, D. and Bigg, T. (1998) *Earth Summit II: Outcomes and Analysis*, London: Earthscan.

Structures of global environmental governance and their political significance

Hass, P.M. (2002) 'UN conferences and constructivist governance of the environment', *Global Governance*, 8(1): 73–91.

Young, O.R. (ed.) (1997) *Global Governance: Drawing Insights from the Environmental Experience*, Cambridge, MA: MIT Press.

International finance mechanisms

Lake, R. (1998) 'Finance for the global environment: the effectiveness of the GEF as the financial mechanism to the Convention on Biological Diversity', *RECIEL*, 7(1): 68–75.

Reed, D. (ed.) (1996) *Structural Adjustment, the Environment, and Sustainable Development*, London: Earthscan.

Reed, D. (1997) 'The environmental legacy of Bretton Woods: the World Bank', in O.R. Young (ed.) *Global Governance: Drawing Insights from the Environmental Experience*, Cambridge, MA: MIT Press, 227–46.

Good governance

Gardiner, R. (2002) 'Governance for sustainable development: outcomes from Johannesburg', presentation to Global Governance 2002, 'Redefining Global Democracy', Montreal, Canada, October.

Radical critique

The Ecologist (1993) *Whose Common Future? Reclaiming the Commons*, London: Earthscan.

Redclift, M. and Woodgate, G. (1997) 'Sustainability and social construction', in M. Redclift and G. Woodgate (eds) *The International Handbook of Environmental Sociology*, Cheltenham: Edward Elgar, 55–70.

Web sites

http://www.earthcharter.org for information about the Earth Charter.

http://www.johannesburgsummit.org for official information about the Johannesburg Summit.

http://www.un.org/geninfo/bp/enviro.html for official UN information.

4 Key global concerns

Climate change and biodiversity management

Key issues

- Mitigation of and adaptation to climate change.
- Climate change and global equity; small island developing states.
- Intergovernmental Panel on Climate Change; UN Framework Convention on Climate Change; Kyoto Protocol.
- Precautionary principle.
- UN Convention on Biological Diversity; Cartagena Protocol on Biosafety; Global Environment Facility.
- National Biodiversity Strategies and Action Plans; Natura 2000; use of plants and animal genetic resources.

The Rio Earth Summit led to two multinational environmental agreements, the UN Framework Convention on Climate Change (UNFCCC) and the Convention on Biological Diversity (CBD). Framework conventions contain loose obligations. The details are worked out through meetings of the conference of the parties (CoP), protocol negotiations, Secretariat work and national implementation. These processes are explored in this chapter.

Climate change is a critical global environmental problem. Exploring climate change means touching upon issues at the core of the sustainable development agenda, including the nature and consequences of economic development and issues of global equity. The problem of climate change is not amenable to simple solutions. Its management has led to the development of a complex and highly controversial global environmental management regime. Climate change also requires cross-cutting policies, which address issues not only in relation to the environment, but across sectors, including transport and energy, and across behavioural areas, including consumption habits and lifestyle patterns.

Biodiversity loss is important because it forces examination of our relationship to nature, including the use of animal and plant genetic resources. Its management, through the CBD, raises issues of North–South relations, the interface between ecology, economy and politics, as well as more general concerns about the function and operation of UN conventions and governance regimes. The UN engagement with climate change is explored first, before attention is turned to the range of issues that have come to be associated with addressing the loss of biodiversity.

Climate change

Causes of climate change

Climate change first attracted the attention of the international policy-making community in the 1980s. There were concerns about the long-term build-up of greenhouse gases in the atmosphere and the effects it could have on the global climate system (Meadowcroft 2002). As a result, in 1988 the UNEP and the World Meteorological Organization established the Intergovernmental Panel on Climate Change (IPCC). Its task was to review scientific knowledge in this area. The IPCC published its First Assessment Report in 1990 (IPCC 1990). So worrying were its findings that they led to the opening of formal international negotiations to deal with climate change (Meadowcroft 2002). Subsequent IPCC reports not only confirmed the reality of climate change, but have also attributed it to human activities.

The build-up of greenhouse gases in the atmosphere is the main cause of climate change. This build-up comes from the burning of fossil fuel and the release of carbon dioxide (CO_2) from industrial production, agriculture and energy generation. This points to the most pervasive and problematic aspect of greenhouse gas emissions: the burning of fossil fuels is an activity that has powered development since the industrial revolution (Meadowcroft 2002). In consequence, dealing with climate change requires, in the short term, the introduction of energy efficiency measures, or processes of ecological modernization, discussed in Chapter 6. Taking a medium-term view, it calls for a technological revolution aimed at the decarbonization of the economy. A longer-term view points towards the reappraisal of industrial development and the shift towards sustainable production and consumption patterns. Here it is useful to make a distinction between:

- action to slow the accumulation of greenhouse gas emissions in the atmosphere in order to forestall further climate change – *mitigation*;
- adapting to climate change that is already under way – *adaptation*;

- removing the structural conditions that cause climate change – *promoting sustainable development*.

As industrialization is the major cause of climate change, it is not surprising to find that the contributions of the industrial countries and the Third World to the problem differ remarkably. The figures are stark: industrialized countries were responsible for four-fifths of CO_2 emissions during the twentieth century and currently account for two-thirds of global emissions. The US *per capita* emissions now stand at over five times the global average, while, at the other end of the scale, many of the poorest countries have *per capita* levels less than 10 per cent of the global average (Meadowcroft 2002: 9). While *per capita* emissions in developing countries are rising, especially in China, they are nevertheless expected to remain well below the US levels for several decades.

The impact of climate change

While the science of climate change is inexact and its findings are subject to different and often conflicting interpretations, there is nevertheless a scientific consensus that its overall impact will be negative. Changing weather patterns are likely to bring increased risk of droughts and flooding, more 'extreme weather events' and sea-level rises. The ecological consequences will include increased risk of species extinction, especially in fragile environments, such as the Arctic and Antarctica. These impacts will not, however, be evenly felt. It is likely that climate change will show regional and local variations.

Climate change will bring major social consequences. These include disruption of agriculture, the erosion of food security, the spread of disease and threats to low-level settlements, with Bangladesh being the most noticeable example. In addition, the ability of countries to take mitigation measures to try to limit the impact and to adjust to climate change problems varies greatly. It is expected that the impact will be most acute in developing countries because they are already more vulnerable and will have less capacity to adapt (Meadowcroft 2002).

Box 4.1 Impact of climate change on small island developing states

The Rio Earth Summit recognized that climate change would present special problems for small island developing states (SIDS), given their ecological and economic fragility. Ecologically, most are archipelagos, with small and dispersed land, possessing rich diversity but relatively few natural resources, and are geographically isolated. The small scale of their economies and the narrow range

---

continued

of products they can produce often make them highly vulnerable to international trading conditions. Agenda 21 pointed to the need for special planning for sustainable development in these circumstances.

The UN recognizes forty-one small island developing states, including the Maldives, Papua New Guinea, the Solomon Islands, Malta, the Bahamas, Haiti and Trinidad and Tobago. In 1994 a UN Global Conference on the Sustainable Development of SIDS was held in Barbados. This resulted in the Barbados Programme of Action, a framework for planning and implementing policies aimed at promoting sustainable development. Fourteen priority areas are identified in the programme. The WSSD also identified several areas for priority action, including in relation to technology transfer and capacity building to improve the sustainable management of coastal areas, implementing programmes of marine and coastal biodiversity protection, implementing sustainable fisheries management, and dealing with problems of waste, pollution, water management and tourism. While small island developing states experience several interrelated environmental pressures, climate change takes priority.

The ecosystems of small island developing states are very sensitive to changes in the environment and they are among the first ecosystems to be severely affected by climate change. Climate change is expected to result in sea-level rises, more extreme weather events and the risk of damage to the rich biodiversity and fragile ecosystems in the islands, such as their coral reefs and mangrove forests. The concern is that climate change will cause irreversible damage to the fragile natural ecosystems of the islands, which in turn will undermine their socio-economic viability.

Exploring small island developing states gives an insight into the connections between the ecological, social and economic dimensions of sustainable development. The problems presented by climate change in particular point to the complex interaction of environmental resources, economic viability and social issues. Even with the Kyoto Protocol in full operation, extreme weather events and sea-level rises are expected. Many argue that the vulnerability of small island developing states is underestimated by the Kyoto Protocol. Ironically, while they have contributed least to greenhouse gas emissions, they will be subjected to some of the most significant adverse effects of climate change.

Source: adapted from Ghina (2003).

The normative dimensions of climate change

Normative issues arise in relation to the causes of climate change, efforts at mitigation and capacity for adaptation. Unveiling these normative issues through the example of climate change helps to show why the promotion of sustainable development has to take account of values and principles, especially those of equity and justice. As Meadowcroft so forcefully argues:

Climate change poses a major challenge for contemporary decision-makers. It involves the design and implementation of policy – with implications across society requiring international co-ordination, and involving substantial costs – in a context of radical uncertainty. It is the most complex environmental issue humankind has had to address. And its resolution is inevitably bound up with issues of distributive justice and the legitimate aspirations of the developing countries.

(Meadowcroft 2002: 33)

Box 4.2 The normative dimensions of climate change

- The industrialized countries bear more historical responsibility for global climate change, an idea known as 'ecological debt'.
- The industrialized world continues to make a higher contribution to climate change.
- The impact of climate change is likely to be unequal, falling especially on small island developing states and low-lying countries.
- The North has more financial and technical ability to mitigate the effects of climate change.

Attention is now turned to how these values and principles have been taken up in the system of international governance that has developed in response to the need to address climate change.

International climate change policy: the UNFCCC

Article 2 of the UNFCCC calls upon member states to 'stabilize greenhouse gas concentrations in the atmosphere at a level that would not disrupt the Earth's climate system'. The UNFCCC has laid down several principles to guide the development of its climate change management regime, all of which are framed with a commitment to promote sustainable development.

To put the principle of common but differentiated responsibilities into effect, the convention makes a distinction between states according to their level of industrialization. Industrialized countries are classified as Annex 1 countries. These are the OECD countries plus the countries with economies in transition in Eastern and Central Europe. Annex 1 countries are held responsible for the largest share of emissions and are obligated to take the lead in mitigation efforts. A further distinction is made within Annex 1 countries in order to isolate the most affluent countries. (These are the Annex 2 countries.) Annex 2 countries, the most prosperous of the industrialized states, have the additional obligation of assisting developing countries to adjust to climate change and to meet their convention obligations.

Box 4.3 The UNFCCC principles

Overriding principle

● Framing policy within the commitment to promote sustainable development.

Normative principle

● Promoting equity, including the obligation of industrialized countries to take the lead in mitigation efforts.

Governance principles

● Making use of the precautionary principle.
● Acceptance of the principle of common but differentiated responsibilities.
● Adhering to the principle of cost effectiveness.

In the period 1992–2005 ten meetings of the parties that signed the UNFCCC were held. These are known as Conferences of the Parties (CoPs) meetings and are aimed at putting flesh on the general agreements outlined in the UNFCCC. CoP meetings have tended to be rather contentious. The 1997 CoP-3 meeting in Kyoto, for example, was particularly divisive for transatlantic environmental relations. The EU arrived at Kyoto armed with a proposal for a binding 15 per cent reduction in emission levels. The US, in contrast, held that such targets were neither technically nor economically feasible. Nevertheless, CoP-3 resulted in the Kyoto Protocol, which sets mandatory limits on emissions by the richer countries, including European countries, the US, Japan and the former Soviet Union. As a result of Kyoto, legally binding targets and timetables have now become an integral part of the implementation of the UNFCCC.

The Kyoto Protocol

The Kyoto Protocol makes use of differentiated obligations, by setting different quantified targets and goals for countries according to their level of economic development. It commits Annex 1 countries to reduce emissions of six greenhouse gases by an average of 5 per cent, relative to 1990 levels. There are, however, different targets set within the Annex 1 group. For example, the EU has a target of 8 per cent reduction, but the so-called 'EU bubble' allows these targets to be redistributed among the member states, so long as they collectively add up to the overall EU allowance. The targets agreed are for the first commitment period, which lasts from 2008 to 2012. New negotiations are required to set the reduction levels for the second commitment period, which begins after 2012. These negotiations are expected to be protracted, as developing countries, including China, will have to be included in a reduction regime. Account will also have to be taken of the fact that industrialized countries, especially Japan, feel that they underestimated the difficulty of meeting targets in the first commitment period and may argue for softer targets in the next period.

There are no reduction targets for developing countries for the first commitment period. Despite the use of differentiated obligations, the developing states, the G-77 plus China, have nevertheless approached climate change in the context of their broader disputes with the industrialized world (Meadowcroft 2002). Developing countries have argued for increased financial flows and technological assistance to shift to cleaner energy sources and help to develop climate adaptation strategies. They are also very reluctant to accept emission reduction targets for the next commitment period, fearing these would undermine development efforts. China is particularly stubborn on this matter. In contrast, the Alliance of Small Island Developing States urges vigorous action, while the OPEC nations, fearing decreases in oil prices, have tried to slow proceedings (Meadowcroft 2002).

The sharpest and most public differences of opinion on the Kyoto Protocol have arisen between the EU and the US. The EU worked for a stringent, legally binding approach to international climate management, while the US has held out against legally binding measures, owing to concern about compliance costs and the possible negative impact on economic activity. Canada, Australia, Norway and Japan initially supported the US defiance. In addition, the US did not ratify the protocol because it lacked binding reduction targets for developing countries, which the US Senate believed gave those countries an unfair advantage in global markets (Bryner 2000). Many US politicians also remain unconvinced of the need for any action in this area, an issue discussed below.

In March 2001 President George W. Bush announced that the US would not implement the Kyoto Protocol. For the Bush administration the protocol has three fundamental flaws:

- It does not oblige developing countries to cut their greenhouse gas emissions.
- It does not allow industrialized countries to comply through investing in reductions in developing countries.
- It is proposing action ahead of further research.

Building upon these objections, the Bush administration dissociated itself from the Kyoto Protocol. The US has not dissociated itself from the UNFCCC, which it has ratified. Its alternative 'Blue Skies' policy, however, amounts to little more than 'business as usual' (Bodansky 2002). 'Most fair-minded assessments have judged the White House's proposed climate change strategy to be wholly inadequate, given the scale of the problem and the magnitude of the US' contributory role' (Cohen and Egelston 2003: 317).

The US decision to abandon the Kyoto Protocol caused a storm of opposition, particularly in Europe and most noticeably from the EU. There is also opposition within the US. In June 2001 the US National Academy of Sciences produced a report, albeit at the request of the Bush administration, which argued that a climate warming trend was evident and that human activity was largely responsible (National Academy of Sciences 2001). In addition, the US Environmental Protection Agency released a report in 2002 describing in graphic detail the negative impact of climate change on sensitive ecosystems in America (US Department of State 2002). There is also strong support outside government for the introduction of policies to deal with climate change, including among the business community, especially the insurance sector. Yet the Bush administration continues to hold to the position that any programme for limiting greenhouse gas emissions must be (1) inclusive of both industrialized and developing countries, (2) predicated upon science and technology, and (3) economically benign. The US is worried that action to deal with climate change could harm its production capacity.

Understanding the US withdrawal

The specific objections that the US has to the Kyoto Protocol can only partly account for the US withdrawal. Several more general explanations can be given, however. First, geo-political rivalry played a big role in the US rejection of the Kyoto Protocol. The failure to include China in the list of Annex 1 countries has proved particularly contentious (Cohen and Egelston 2003: 320). Most energy analysts project that the Chinese will become the foremost producer of greenhouse gas in the period 2010–25. As China develops economically, it is increasing its presence on the international stage and now challenges US interests on a broad range of issues. In this context, the US rejection of the Kyoto Protocol can be seen as part of a strategy of ensuring China gets no concessions that would enhance further its comparative economic advantage (Cohen and Egelston 2003). Second, the US rejection of the Kyoto Protocol should not be seen in isolation, as it has rejected several other international initiatives, including a treaty to ban land mines. It has also adopted a solitary position on a number of other major issues. The rejection of the Kyoto Protocol is as much part of this wider strategy as it is a reflection of an administration that gives very low priority to environmental matters.

The US position on Kyoto can also be explained by reference to more general matters of economic ideology. The Bush administration endorses the belief that economic growth and deregulation are prerequisites of innovation, especially in the industrial sector. In contrast, regulation is seen in Europe as offering incentives to invest in innovation and to experiment with alternative approaches, as discussed in the analysis of ecological modernization in Chapter 6. Ecological modernization has not taken hold in the US (Baker and McCormick 2004). This, combined with an alliance between the White House administration and an array of declining industries, such as oil and coal, insulates US policy makers from challenging new perspectives in relation to reconstructing the relationship between economy, ecology and society (Leggett 2001).

> [C]ontinued allegiance in the USA to an increasingly anachronistic economic-environmental model is problematic. At a time when environmentally attentive publics elsewhere in the world are encouraging their governments to embrace progressive environmental strategies that enhance innovation, the Bush administration remains unyielding in its view that economic and environmental objectives are irreconcilable. This obsolete notion may have been credible during a prior industrial era before consumers and investors began to inculcate a modern sense of environmental responsibility.
>
> (Cohen and Egelston 2003: 325)

This argument provides an explanation of why the agenda of sustainable development has not taken hold in the US (Baker and McCormick 2004). Within the US, however, there is a creative array of sustainable development initiatives which is especially noticeable at the community level.

One major source of controversy between the US and parties to the Kyoto Protocol centres on the status of our knowledge about climate change. The US argues that further scientific clarity is needed before action is taken on the matter. Waiting would also bring the advantage that the more technologically developed society of the future would be better placed to take action to address the issue. Others disagree, pointing to the substantial body of scientific consensus on human-induced climate change that already exists, arguing that deferring action may result in irreversible damage, and explaining that the sooner the problem is addressed the longer the time available to adapt to climate change and to begin the transition to decarbonization of energy systems.

At the heart of the dispute over the status of our knowledge about climate change there is a difference of opinion over the application of the precautionary principle (Box 4.4).

Box 4.4 The precautionary principle

In its simplest form, the precautionary principle, as a tool of risk management, holds that, in the face of scientific uncertainty, policy makers should err on the side of safety. The principle is of German origin, *Vorsorge Prinzip*, and has shaped German environmental law since the 1980s. The precautionary principle is particularly important for understanding the differences between the approach of the EU and US to climate change.

The principle has been in use in international environmental policy for over two decades. It was first recognized in the 1982 UN World Charter for Nature and has subsequently been incorporated into several international environmental conventions. It is in the 1992 Rio Declaration and the UNFCCC refers to the precautionary approach, where Article 3 (Principles) states that 'The Parties should take precautionary measures to anticipate, prevent or minimise the causes of climate change and mitigate its adverse effects. Where there are threats of serious or irreversible damage, lack of full scientific certainty should not be used as a reason for postponing such measures.'

Since the 1980s the precautionary principle has been progressively consolidated in international environmental management regimes, making it 'a full-fledged and general principle of international environmental law' (CEC 2000a: 10). This despite growing US opposition to the use of the principle. The US fears that the principle could be used for (trade) protectionism purposes, especially by the EU.

The EU Commission also wishes to evoke the principle to manage risk in the longer run and for the well-being of future generations (CEC 2000a). Consideration of the inter-generational dimensions of environmental management is one of the distinctive features of sustainable development. Here we can see how the commitment to the promotion of sustainable development influences the Commission's interpretations of key related policy principles.

Source: Baker (2005a).

Consequences of US withdrawal

The US withdrawal from the Kyoto Protocol has had a major impact upon international efforts to address climate change. It challenges countries to continue their commitment despite the fact that the US is not introducing measured policies to address climate change. The EU, for example, sees this asymmetry as potentially dangerous for European industrial competitiveness (CEC 1997a). There is also concern that the US could act as a free rider, benefiting from costly efforts undertaken by others without having to incur those costs itself (CEC 1997b). The US withdrawal, and its failure to develop an alternative, proactive climate change policy, force the EU to make a choice between environmental values and the priority given to economic considerations.

In response to this direct challenge, the EU has remained determined to show itself a global leader in the area of climate change (Haigh 1996), as it did in the early years of the IPCC (1988–91) (Bretherton and Vogler 1999). Its stance has given the EU a heightened international role in climate change management. This was seen at the resumed CoP-6 *bis* negotiations in Bonn in 2001, which, to the great surprise of the US, reached agreement on the implementation of the first commitment period of the Kyoto Protocol (Vrolijk 2002). The leadership role of the EU was also evident at the CoP-7 meeting in Marrakech, resulting in the Marrakech Accord. Paradoxically, it would appear that the US disavowal of Kyoto helped to galvanize the resolve of other states and the EU to reach agreement that would allow the Kyoto Protocol to come into force (Meadowcroft 2002). Yet global environmental leadership comes new to the EU, as was evident in its bungling behaviour at the CoP-8 in Delhi. There is still much at stake as the EU struggles to ensure that its actions contribute to, rather than undermine, the desire to come of age on the international political stage (Grubb and Gupta 2000). The recent shift in the US approach, from a rather benign to a more hostile position on Kyoto, adds to this challenge (Ott 2003). The Johannesburg WSSD made major progress, however, when Canada, Mexico, China and Russia declared their commitment to ratifying the protocol. Russia's commitment was particularly significant, since Russia contributes 17 per cent of the CO_2 emissions from all industrialized countries.

Kyoto without America: moving ahead with the Kyoto Mechanisms

Many, including the EU, see the use of policy instruments, as opposed to sole reliance upon command-and-control legislation, as essential to the promotion of sustainable development. This is because their use stimulates the involvement of a wider range of actors in the process. It is also justified on the grounds of efficiency.

Policy instruments using incentive mechanisms have found their way into the policy process in a variety of areas in a number of countries. Examples include the establishment of individual transferable quotas in the fisheries in Australia, Iceland and New Zealand, and the introduction of tradable emissions permits in an effort to curtail sulphur dioxide emissions in the US (Young 2003). Use is made of a similar range of instruments to meet the targets of the first commitment period of the Kyoto Protocol. These include emissions trading, Joint Implementation and the Clean Development Mechanism. The Marrakech Accord added the use of carbon sequestration measures to the set of policy tools. These instruments are known collectively as the Kyoto Mechanisms (Box 4.5).

Box 4.5 The Kyoto Mechanisms

- *Emissions trading*: emission permits are allocated to Annex 1 member states. They devise methods to distribute these permits to emitters of greenhouse gases under their jurisdiction. Emissions trading allows countries that can achieve low-cost abatement to sell emission entitlements to other countries that are having more difficulty meeting their targets, thus reducing the overall cost of international compliance.
- *Joint Implementation*: this allows states to gain credit for emission reductions that they helped to achieve in another country within the Annex 1 group. For example, they could undertake a particular project, such as helping to replace an outmoded power plant with a more carbon-efficient alternative, provided it is not a nuclear plant.
- *Clean Development Mechanism*: this is where an Annex 1 country helps a non-Annex 1 country and can gain credit for emissions reduction, for example by helping with reforestation projects.
- *Carbon sequestration*: this is the long-term storage of carbon in geological formations, such as oil wells.

The protocol also allows 'land use, land-use change and forestry' (LULUCF) activities to be included, as these can, for example, increase long-term carbon storage in 'natural sinks' which can draw carbon down from the atmosphere and store it in forests. However, this use has been criticized by many environmentalists, who remain unhappy about carbon sequestration, fearing the unknown effects of interference with nature over geological time. There is also concern that, because the use of LULUCF is counted against Kyoto targets, it will not reduce the net level of emissions.

Many major problems surround the use of the Kyoto Mechanisms. Daunting problems exist with respect to conflicts over the initial allocation of the Kyoto targets, with respect to non-participation, enforcement and monitoring, and in relation to the exploitation of numerous loopholes in the climate change agreements. Some countries, for example, intend to use the inclusion of sequestration in the list of Kyoto Mechanisms to argue that they need not take any steps to address climate change. Canada, for example, holds that the sequestration of carbon in its forests is sufficient to excuse Canada from making any substantial reductions in greenhouse gas emissions (Young 2003).

Emissions trading has been subject to particularly strong criticism. For many environmentalists, emission trading breaches the principle of environmental integrity. It involves allocating the right to pollute, via emission permits, and turning pollution sources into tradable commodities. In addition, the allocation of emission permits at the international level has had unintended consequences.

Choosing 1990 as the base year gave approximately 25 per cent of the permits to the US, an inappropriately large proportion to the countries in transition and relatively few to developing countries. This procedure actually rewards polluters for their past antisocial behaviour (Young 2003). It also allows some countries to reap large financial gains from emissions trading. The Russian Federation, for example, hopes that, having signed the Kyoto Protocol, it will benefit financially from the sale, especially to the EU, of what has become known as 'hot air' – that is, the excess emission allowance of Russia and other former communist countries. This excess appeared when the collapse of their industrial production resulted in them not needing the amount of emissions that they were granted.

Limitations of Kyoto

The IPCC has indicated that, in order to stabilize concentrations of greenhouse gas in the atmosphere, emissions have to be reduced radically. Given its low emission targets, the protocol is scorned as environmentally ineffective: its targets are set so low that, even if fully implemented, which is doubtful, they will not halt the rise in greenhouse gas emissions, let alone address climate change. Rapidly rising emissions from developing countries will more than neutralize whatever abatement is reached, even if proportionate reductions are agreed in the next commitment period. However, as Meadowcroft has argued, the Kyoto Protocol 'was never intended as a comprehensive solution to climate change'. Instead the protocol has to be:

> understood as part of a long-term process to create global institutions to stabilise atmospheric concentrations of greenhouse gases. It is a first step that commits Annex 1 countries to modest initial reductions, and can serve as a bridge towards more substantial future cuts, and the extension of the pool of participating countries, in subsequent commitment periods.
>
> (Meadowcroft 2002: 15)

Biodiversity

The context

Biological diversity, or biodiversity, and ecosystems provide 'ecological services', such as CO_2 absorption, clean water, plant pollination by insects and nitrogen fixation. If these ecosystems are disrupted, the richness and variety of the natural world are reduced. Human societies may encounter newly emerging diseases and suffer losses and damage to forest and marine resources. In short, biodiversity is a crucial indicator of planetary health (Iles 2003).

Biodiversity loss results from the encroachment of people on ecosystems and, as such, it is a consequence of human interaction with the natural world. What is problematic is that the rate of biodiversity loss is disrupting the replenishment capacity of natural ecosystems. Biodiversity, in terms of ecosystems, species and genetic diversity, is being destroyed at an alarming rate. Forests, which hold the highest number of species of all terrestrial ecosystems, are under severe threat, especially in subtropical and tropical areas. Indonesia, for example, has lost almost a quarter of its 1985 forest cover. An estimated 27 per cent of the earth's coral reefs have been severely damaged. According to the World Conservation Union's Red List of Threatened Species, an estimated 24 per cent of all mammals, 12 per cent of birds, 25 per cent of reptiles, 20 per cent of amphibians, 30 per cent of fish and 16 per cent of conifers are threatened. There is also evidence of severe genetic erosion of cultivated plants and animals. According to the Food and Agriculture Organization (FAO) of the United Nations, some of the leading 'provider countries' of crop plants, such as wheat and maize, have lost more than 80 per cent of their plant varieties. Agricultural systems based on industrial monoculture may be highly susceptible to plant disease, climate change and ecological shifts. 'The rate of biodiversity loss is increasing at an unprecedented rate, threatening the very existence of life' (Secretariat CBD 2002: 304).

Biodiversity is under threat from several sources. First, biodiversity loss is caused by the interplay between particular economic sectors (such as agriculture, energy and transport) and individual ecosystems. As such, biodiversity protection requires effective policies to integrate environmental considerations into other policy areas, a process known as environmental policy integration, discussed in Chapter 6. Biodiversity is also under threat from global problems, such as greenhouse gas emissions, which affect all ecosystems. As a result, there are now overlaps between biodiversity protection and the objectives of the growing number of international environmental management regimes, such as the UNFCCC, and their associated MEAs.

Since the mid-1990s, several controversial interfaces have developed between biodiversity and the spheres of politics and commerce. The development of the biotechnology industry in particular has given rise to concern about biosafety. There is also concern about the use of 'Genetic Use Restriction Technologies', including so-called 'terminator technology', a technology that induces sterility in the second generation of crops. This can lock Third World farmers into structures of dependence as they are forced to buy seed from biotechnology companies at the start of each sowing season. There is also anxiety about access to genetic resources, about who has 'ownership' of genetic resources and who has the legitimate right to their use or to negotiate access to them. This, in turn, is related to unease about the international system of intellectual property rights. The growing commercial significance of biodiversity resources is also raising concern about bio-piracy among Third World countries as biotechnological and pharmaceutical firms, particularly from the industrialized world, seek out the biological resources of developing countries, in the hope of identifying new candidates for commercial exploitation. This has the potential to threaten biodiversity through commercial over-exploitation and the disruption of traditional patterns of biological resource use (Iles 2003). This disquiet is occurring against a background of growing social controversy over biotechnology, including concern over the potential negative impact on biodiversity of the release into the environment of genetically modified organisms (GMOs).

The normative principles of the CBD: linking biodiversity and sustainable development

Like its sister convention, the UNFCCC, the CBD is based upon an articulated set of normative and governance principles. While the convention seeks to promote a sense of shared responsibility for the protection of biodiversity at the global level, it nevertheless reaffirms the role of the state as the key actor in biodiversity protection. In keeping with UN traditions, it upholds the principle of subsidiarity, laying down that 'the authority to determine access to genetic resources rests with the national governments and is subject to national legislation' (Article 15). The CBD also explicitly draws upon the precautionary principle, stating that 'Where there are threats of serious or irreversible damage, lack of full scientific certainty shall not be used as a reason for postponing cost-effective measures to prevent environmental harm' (Article 15). The CBD also adopts an ecosystem approach, a holistic approach to conservation that has become a central tenet of the convention.

The CBD has three main objectives, built upon an explicit commitment to promote sustainable development:

- the conservation of biological diversity;
- the sustainable use of its resources;
- the equitable sharing of the benefits arising from the use of genetic resources.

These objectives go well beyond narrowly defined conservation measures. In fact the CBD is based on the belief that 'addressing the threats to biodiversity requires immediate and long-term fundamental changes in the ways that resources are used and benefits are distributed' (Secretariat CBD 2002: 305). Its concerns range from ecosystem protection to the exploitation of genetic resources, from conservation to questions of environmental and social justice, from commerce to scientific knowledge, and from the allocation of rights to the apportionment of responsibility (Le Prestre 2002). The convention makes it clear that the conservation of biological diversity requires addressing the *causes* of biodiversity loss and the political, economic and social processes that foster it. This means addressing issues such as property rights, trade patterns, inequitable social relations and unsustainable patterns of economic development and resource consumption. Given the scope of the issues it addresses and the links that it makes between the ecological, economic and social dimensions, it has been argued that the CBD is 'the first truly and for the moment the foremost sustainable development treaty' (Le Prestre 2002: 270).

However, these very characteristics are also the source of its weakness. They have also led to tensions among the signatories to the convention. The group of developing countries (G-77) support the CBD precisely because of the links it makes between its three basic goals of conservation, sustainable use and benefit sharing. Others, particularly the US, would prefer to see the three objectives of the convention decoupled, because these links move the convention too deep into the political arena and thus into the muddy waters of North–South relations. Similarly, there are concerns that, while the CBD upholds the principles of equity and fair sharing of genetic resources and their benefits, a number of important issues remain to be addressed. These include the need to ensure the participation of indigenous groups, and to respect different cultural values as they relate to the use of plant and animal genetic resources. Account also has to be taken of different understandings of rights, as some traditions see them as vested in the individual while others understand rights only in relation to the community, tribe or group.

As the convention can be characterized as wide-ranging, ambitious and deeply political (Le Prestre 2002), it is not surprising to find that, from its birth, the CBD has been plagued with problems. The US refused to add its signature to the convention, as it could not accept the clause dealing with intellectual property rights. Japan, as well as two EU member states (the UK and France), expressed similar concern, fearing competitive disadvantages for their growing biotechnology sector (Baker 2003). Later, at the New York Summit in 1997, the on-going dispute over the regulation of biotechnology and the protection of intellectual property rights led to virtual stalemate over biodiversity.

The CBD joined a crowded field of multinational and regional environmental and development agreements. This made it all the more important to clarify the boundaries of the activities of the CBD and to specify what range of issues is central to its goals. Given the broad scope of the CBD and the ways in which it links biodiversity conservation to the broader task of promoting sustainable development, this has not proved to be an easy task. As a result, coordination with existing legislation and MEAs, such as the Convention on International Trade in Endangered Species of Wild Fauna and Flora (CITES), has proved difficult. Similarly, while it is clear that there is a relationship between the Convention and Agreement on Trade-related Aspects of Intellectual Property Rights (TRIPS) and the CBD, it is clearly an enormous task to reconcile biodiversity protection, environmental protection, human welfare, trade liberalization and property rights. It is also foolish to assume *a priori* that there are no conflicts between the goals of protecting ecosystems, species and biodiversity and the promotion of sustainable development.

The Biosafety Protocol

As is the case within the international climate change management regime, regular CoPs take place that aim to establish programmes of work and set priorities to implement the CBD. By and large, negotiations among the CoPs have been difficult, not least because of the wide scope of the convention and the highly contentious issues that have risen around it. There were calls made at the CoP-3 in 1996 to streamline the work of the CoP, and it was not until the CoP-6 meeting in 2002 that a draft Strategic Plan for implementing the CBD was developed. Arguably, the most successful work to emerge from the CBD is the agreement reached in 2000 on biosafety, the Cartagena Protocol on Biosafety (Box 4.6).

Box 4.6 The Cartagena Protocol on Biosafety

The Cartagena Protocol on Biosafety (the Biosafety Protocol), agreed in 2000, is an MEA which provides legally binding measures to promote and monitor the transfer, handling and use of 'living modified organisms' which are the result of genetic engineering. Living modified organisms can be distinguished from GMOs, in that they are living, and thus can escape into the environment and grow. Anything that is no longer living, but has been altered genetically, is classified as a GMO and is not subject to the protocol.

The protocol is designed to protect a nation's domestic environment from the accidental release and spread of GMOs. It is based on two principles: the precautionary principle and the principle of prior informed consent of receiving countries.

The objectives of the Cartagena Protocol on Biosafety are to contribute to the 'safe transfer, handling, and use of living modified organisms resulting from modern biotechnology that may have adverse effects on the conservation and sustainable use of biological diversity, taking into account risks to human health, and specifically focusing on transboundary movements' (Protocol, Art. 1, 'Objectives').

---

continued

The protocol is not intended to restrict trade in living modified organisms. On the contrary, it is premised on the belief that 'trade and environment agreements should be mutually supportive with a view to achieving sustainable development'. It represents a good example of an MEA with trade implications, making it of particular relevance to the WTO, as discussed in Chapter 6. The US administration of President George W. Bush criticizes the Biosafety Protocol, arguing that it could be used to control international trade, particularly in agricultural products. Others argue that it will do the opposite: by developing a management and regulatory regime for living modified organisms, the protocol could encourage trade in biotechnological products, despite the fact that little is known about their impact on biodiversity and despite growing public disquiet.

The adoption of the Cartagena Protocol has had a major impact on the development of the CBD. On the positive side, the success in negotiating the protocol has given rise to a 'feel-good factor', and has helped channel increased resources into the convention. On the other hand, there is a danger that the management of the Biosafety Protocol will become a central activity, diverting attention and resources from other dimensions of the convention.

Source: adapted from Anderson (2002).

Funding biodiversity protection: the role of the Global Environment Facility

Biodiversity protection is financed through the Global Environment Facility (GEF) and is its largest portfolio. By 2001 the GEF had supported over 130 developing countries to construct national biodiversity strategies. However, at least 40 per cent of biodiversity funding from the GEF is in the form of borrowing. This increases the risk that developing countries may escalate their debt load in order to fund biodiversity protection. The irony is that the need to finance debt repayments may lead, in turn, to pressure for unsustainable exploitation of the very genetic and biological resources that the CBD is designed to protect (Iles 2003).

GEF funds are not intended to meet the total cost of achieving the CBD's objectives, only the so-called 'incremental costs' – that is, that part of the total cost that yields global benefits and will not be incurred by a country in the course of its 'normal' development. This makes it very difficult for the GEF to deal with the underlying causes of biodiversity loss. There is also an on-going problem with respect to the replenishment of GEF funds. With the US in arrears, and other industrialized countries refusing to increase their own contributions purely to make good the shortfall caused by the US, the GEF operates under a very restrictive budget. There are also difficulties of a more political nature that have soured relations between the CoP and the GEF. Because of its close relationship with the World Bank, the GEF is viewed with suspicion by both Third World countries and environmental NGOs, as discussed in Chapter 7.

Implementing the CBD at the national level

The implementation of the CBD crucially depends on the creation of National Biodiversity Strategies and Action Plans. Several countries have drawn up National Strategies: the UK and Sweden as early as 1994, Canada, Japan and Vietnam closely following in 1995. Most point to a lack of basic knowledge about their country's biodiversity and, especially in developing countries, lack of capacity to undertake research in this area. National Strategies often call for the creation of protected sites and parks, but such sites are often established in sparsely populated areas, to minimize conflict over land use, and not necessarily in areas with the highest biodiversity or in the most urgent need of protection. In addition, the protected areas do not necessarily cover a representative range of ecosystems, habitat species and genetic diversity (Baker 2003). In some cases, sites were established without the introduction of regulations to allow indigenous and local communities to continue their traditional use of natural resources. This separates local people from their environment, and can result in lack of support or even open opposition from the inhabitants of an area to the designation of their land as a nature park, on the grounds that the biodiversity protection is another example of the closure of the commons. 'In the guise of conserving biodiversity, parks can be created that deprive people of their land and livelihoods, and are open to foreign corporations for bio-prospecting without sharing the benefits' (Iles 2003: 232).

In addition, many of these sites are not sufficiently managed and continue to allow land-use practices that run counter to the objectives of the site. Thus some protected areas have been portrayed as mere 'paper parks' (Herkenrath 2002).

Third World countries face several difficulties in the implementation of the CBD and in particular in responding to the requirement to construct National Strategies. The GEF has assisted more than 125 developing and transition countries in producing their strategies. Those that have developed a National Strategy, however, often find that associated legislation is missing, such as endangered species legislation. Implementation in developing countries also requires both technology transfer and financial assistance. However, such capacity enhancement is not a neutral transfer process. There is a real danger that the transfer of funds, technology and knowledge from industrial countries to Third World countries may be conditional upon opening up access to genetic and biological material. It can also encourage developing countries to pursue high-technology, modernist, and intensive development based on the harvest of biological material for pharmaceutical, industrial and agricultural uses. This could represent little more than an updated version of the development model that the World Bank and other international donors promoted in the Third World from the 1960s to the 1990s. This development path is now seen as having contributed to debt, poverty and ultimately to unsustainable patterns of resource use, as indebted countries were over-harvesting their natural resources to support their international debt repayments.

This reflects a more general criticism of the work of the CoPs, namely that meetings have devoted more attention to reaching agreement on access to genetic resources and biotechnology, as well as transfers of funds in order to facilitate the harvesting of such material, and have paid less attention to the development of guidelines for the sustainable use of the resources and for equitable benefit sharing (Iles 2003). This promotes a weak sustainable development position, as it values biodiversity only in relation to commercial use.

Significance of the CBD

The chief significance of the CBD is that is has helped to forge a link between biodiversity protection and the promotion of sustainable development. Biodiversity preservation is now an integral part of the construction of our sustainable future. Because of the CBD, biodiversity is redefined in social and economic terms, and no longer seen as a mere technical, scientific issue.

The CBD is a very difficult convention to implement. Its complexity and scope, its relative lack of public visibility, especially as compared with climate change, its political ramifications and the underdeveloped nature of its key tools represent significant challenges (Le Prestre 2002). The obstacles to the implementation of the CBD are also of a structural nature. Consumption patterns in high-consumption societies and international trade rules, for example, need to undergo fundamental change if the preservation of biodiversity is to be assured. There is also a strong sense in which the protection of biodiversity is reliant upon the promotion of a new relationship with nature, one that is based upon recognition of the intrinsic value of biodiversity, rather than merely upon the dominant utilitarian argument that biodiversity should be protected in so far as it is of use to us. The latter view makes biodiversity protection vulnerable to shifting cultural, political and economic perceptions and values.

The CBD is an integral part of, and reflects the UNCED understanding of, sustainable development: it does not posit conservation as the pillar of the relationship between society and nature. Rather, it affirms the primacy of social and economic development, while aiming to couple that development with biodiversity protection. In so far as the CBD can stimulate structural change, it has the potential to contribute to the promotion of strong forms of sustainable development. Work to date among the contracting parties to the convention would suggest, however, that the CBD is more likely to promote a weaker form of sustainable development, dominated by utilitarian, particularly commercial, views of nature and its biodiversity resources.

Conclusion

In this chapter two UN conventions have been explored, the first designed to address climate change and the second focusing on the maintenance of bio-diversity. Both are highly ambitious conventions, because they make direct links with the need to promote sustainable development. By exploring the range of issues that have come to be associated with these conventions, insight is gained into the complex, dynamic and uncertain nature of the cross-cutting tasks involved in promoting sustainable development.

Their location within the UN system frames both conventions within a sustainable development agenda. However, it also means that very complex institutional structures, negotiation processes and sets of agreements surround both conventions. In addition, it has brought a legacy of problems to both conventions, allowing Third World countries to approach both climate change and biodiversity maintenance within the context of their broader disputes with the industrialized world, while embroiling both conventions in the controversial relationship between the US and the UN. The attitude of the US has had substantive and very negative outcomes for both conventions.

Conventions set out broad areas of agreement, making it necessary for CoPs to flesh out the details, agree targets and timetables, and establish reporting and monitoring mechanisms. However, CoP meetings have presented ideal opportunities for the articulation of narrow self-interest, aimed at achieving short-term, often commercial, gains. This is very different from the democratic, participatory processes envisaged by Brundtland.

A recurring theme in the discussion was the need to address the structural causes of both climate change and biodiversity loss. Intensive resource use in the high-consumption societies, as well as the structures of international trade, need to undergo fundamental change if the preservation of biodiversity is to be assured and if climate change is to be addressed. Neither convention is capable of bringing about such change.

However, what the conventions have been able to do is show the links between the promotion of sustainable development and the maintenance of the planet's ecological health. There is now clearer recognition of the central role played by the ecological and planetary systems in the construction of our sustainable future. Rather than seeing these as mere technical, scientific issues, the conventions have helped redefine both biodiversity and climate change in social, economic and political terms.

Summary points

- Climate change will bring about major social consequences. The ability of countries to take measures to try to limit the impact and to adjust to climate change problems varies greatly.
- Several of the normative principles associated with the promotion of sustainable development are relevant to the issue of climate change.
- The 1997 Kyoto Protocol is part of a long-term process to create global institutions to stabilize atmospheric concentrations of greenhouse gases.
- The EU has shown itself a global leader in the area of climate change. The US has withdrawn from its obligation to address the problem.
- Several factors can account for the US withdrawal: geo-political rivalry, outmoded understanding of the relationship between environment and economy, disputes over the precautionary principle.
- The CBD has three principal objectives: (1) the conservation of biological diversity; (2) the sustainable use of resources; (3) the equitable sharing of the benefits arising from the use of genetic resources.
- The CBD has to deal with several controversial interfaces between bio-diversity and the spheres of politics and commerce.
- The most successful work to emerge from the CBD is the 2000 Cartagena Protocol on Biosafety, which has led to National Biodiversity Strategies and Action Plans.
- The CBD reflects UNCED understanding of sustainable development: it affirms the primacy of social and economic development, while aiming to couple that development with biodiversity protection.
- Both conventions are built upon recognition of the central role played by the environment and planetary systems in the construction of our sustainable future. Rather than seeing them as merely technical, scientific issues, the conventions have helped define both biodiversity and climate change in social, economic and political terms.

Further reading

Analysis of the politics of climate change

Cohen, M.J. and Egelston, A. (2003) 'The Bush administration and climate change: prospects for an effective policy response', *Journal of Environmental Policy and Planning*, 5: 315–31.

Meadowcroft, J. (2002) *The Next Step: A Climate Change Briefing for European Decision-Makers*, Policy Paper 02/13, Florence: European University Institute.

Convention on Biodiversity

Herkenrath, P. (2002) 'The implementation of the Convention on Biological Diversity: a non-government perspective ten years on', *RECIEL*, 11: 29–37.

Iles, A. (2003) 'Rethinking differential obligations: equity under the Biodiversity Convention', *Leiden Journal of International Law*, 16: 217–51.

Le Prestre, P.G. (2002) 'The CBD at ten: the long road to effectiveness', *Journal of International Wildlife Law and Policy*, 5: 269–85.

Mang, J. (ed.) (2000) *Root Cause of Biodiversity Loss*, London: Earthscan.

Global environmental governance

Young, O.R. (2003) 'Environmental governance: the role of institutions in causing and confronting environmental problems', *International Environmental Agreements: Politics, Law and Economics*, 3: 337–93.

Web resources

CBD: for official information from the CSD Secretariat see http://www.biodiv.org. For information on funding see http://www.gefweb.org.

Climate change: www.climatenetwork.org, the website of the NGO Climate Action Network; http://unfccc.int, the official UNFCCC site.

5 The local level

LA21 and public participation

Key issues

● Participation, democracy and civil society; Aarhus Convention.
● Local authorities, planning and the spatial dimension.
● Agenda 21 and Local Agenda 21.
● Ålborg Charter and Sustainable Cities campaigns.

This chapter examines the promotion of sustainable development at the local level through Agenda 21, the action plan for sustainable development adopted at the Earth Summit in 1992. Its particular focus is on Local Agenda 21 (LA21), a scale at which action can have both immediate and direct effect. Because LA21 gives local authorities a key role, this provides an ideal opportunity to explore the relationship between planning and the promotion of sustainable development. The spatial dimension of sustainable development, including in relation to urban design, land use and transport, comes into sharp focus, particularly in cities. By looking at LA21, insight is also gained into how the launch of new participatory processes based on good governance principles has helped local actors identify what is needed at the local level. While attention is focused at the local level, account has also to be taken of how broader processes, across the different levels of global, international, national and local governance, act as facilitators or as impediments to change.

Agenda 21

The Agenda 21 document begins with an astute analysis of the causes of unsustainable development. It points to the non-sustainable patterns of production and consumption in wealthy countries as the most significant cause of environmental degradation. It then sets out a blueprint for working towards development that is socially, economically and environmentally sustainable. Its solutions promote development that does not compromise the natural resource base and the ability of future generations to sustain themselves. Solutions also emphasize the importance of citizen participation (http://www.prosus.uio.no/english/local/la21/index.htm, accessed 11 May 2004).

Agenda 21 is organized into four sections, and its forty chapters address the major areas in which political action is needed (Box 5.1). Agenda 21 emphasizes the importance of creating adequate knowledge and institutions, including through education and through the development of human resources. This is known as 'capacity building' in UN jargon. Missing from Agenda 21 are discussions of several important but highly controversial issues, known as the 'black holes' (Dresner 2002: 42); these include population, international debt and militarism. For example, Chapter 5, dealing with population, does not refer to contraceptives, at the insistence of the Vatican and the Philippines (Dresner 2002).

Box 5.1 Main areas of action outlined in Agenda 21

- *Section I, Social and economic development* (chapters 2–8): international cooperation; combating poverty; changing consumption patterns; addressing population growth; protecting human health; promoting sustainable human settlements; ensuring environmental policy integration.
- *Section II, Conservation and management of resources for development* (chapters 9–22): protection of the atmosphere; introducing integrated land planning and management; combating deforestation; ensuring sustainable management of fragile ecosystems; combating desertification and drought; sustainable mountain development; promoting sustainable agriculture and rural development; conservation of biological diversity; environmentally sound management of biotechnology; protection of oceans and seas; protection of the quality and supply of freshwater resources; ensuring environmentally sound management of toxic chemicals and all types of waste.
- *Section III, Strengthening the role of major groups* (chapters 23–32): ensuring the participation in actions and plans of women, youth, indigenous peoples, NGOs, local authorities, workers and trade unions, business and industry, the scientific community and farmers.
- *Section IV, Means of implementation* (chapters 33–40): financial resources, including replenishment of ODA and dealing with debt; technology transfer; cooperation and capacity building; using science for sustainable development; promoting education, public awareness and training; international cooperation and information sharing for capacity building; enhancing international institutional arrangements; strengthening international legal instruments.

Source: adapted from Koch and Grubb (1993).

Section III of Agenda 21 is devoted to the involvement of governmental agents, social groups and the business community. It places considerable emphasis on developing new forms of democratic governance and on enhancing popular participation. This is based on the belief that effective and legitimate change requires active involvement by interest groups and other organizations. The idea of giving local communities a say in shaping the formulation and implementation of policies is in keeping with the beliefs of many Green activists, who give local knowledge a privileged position as part of the pursuit of sustainable development.

In addition, the promotion of sustainable development is seen as requiring new forms of social learning. These can include learning how to engage in constructive dialogue with others, for example how to take account of the interests of others and not just to uphold one's own narrow concerns, and how to envisage collectively the elements needed to construct a sustainable future. As will be seen below, this learning forms an important part of LA21 activity and is closely related to the idea of using social capital as a tool in the promotion of sustainable development. However, in keeping with the UN system of governance, national governments have overall responsibility for the implementation of Agenda 21.

Local Agenda 21

Chapter 28 of Agenda 21 is devoted to the role of local political authorities in the introduction of comprehensive planning processes aimed at promoting sustainable development within their locality. This activity has come to be known as Local Agenda 21, or LA21. More specifically, LA21 refers to the general goals set out in Chapter 28 of Agenda 21.

Local authorities are singled out because they have specific and significant environmental management functions and responsibilities. These include:

- developing and maintaining local economic, social and environmental infrastructure;
- overseeing planning and regulations;
- implementing national environmental policies and regulations;
- establishing local environmental policies and regulations.

As the level of government closest to the people, local authorities also play a vital role in educating and mobilizing the public. The role of local authorities in planning has come to be seen as particularly important. Their control over land-use planning, for example, shapes the form of urban development within a particular area. This, in turn, determines whether a city is compact or sprawled and can influence the design of new buildings. The form of urban development can then influence energy use, thus making planning an important tool in combating climate change.

Local authorities are expected to act as catalysts in the start-up of LA21 initiatives and, subsequently, as facilitators, ensuring the participation of a wide range of actors, drawn from within their local community. This participation is intended to lead to the formation and subsequent implementation of long-term strategies that focus on the local level. While local authorities are the main facilitators, there is also a strong role envisaged for national governments, including launching a national campaign directed at encouraging local authorities to act. It is intended that LA21 will be stimulated by and, in turn, contribute to the development, structures and processes of international cooperation.

> LA 21 is a participatory, multistakeholder process to achieve the goals of Agenda 21 at the local level through the preparation and implementation of a long-term, strategic plan that addresses priority local sustainable development concerns.

(ICLEI 2002: 3)

Putting LA21 into practice

Chapter 28 sets detailed targets and timetables for local authority action (Box 5.2). Institutionally, LA21 is supported by the International Council for Local Environmental Initiatives (ICLEI), an international association of local governments founded in 1990 with the help of UNEP. It influenced the writing of Chapter 28 and has since played a key role in stimulating the development of LA21. It acts as a co-coordinator and facilitator, especially at the start-up stage of LA21 activities, as a clearing house for information, as a major conduit for the transfer of expertise and best practice, especially between the North and the South, as well as an organization representing local authorities engaged in LA21 activities at the international level.

Box 5.2 Targets and timetables for LA21

- *By 1996* local authorities in each country should have undertaken a consultative process with their populations and achieved a consensus on an LA21 plan for their community.
- *By 1993* the international community should have initiated a consultative process aimed at increasing cooperation between local authorities.

---

continued

- *By 1994* representatives of associations of cities and other local authorities should have increased levels of cooperation and coordination with the goal of enhancing the exchange of information and experiences among local authorities.
- All local authorities in each country should ensure that women and youth are represented in decision making, planning and implementation processes.

Source: adapted from Koch and Grubb (1993: 136–42, 147–51).

One of the major contributions of ICLEI has been to develop guiding principles for LA21 actions. These constitute rules of good governance practice (Box 5.3). More specifically, establishing an LA21 plan for a local authority area typically involves six key steps (Box 5.4). Despite the detailed nature of these objectives, Chapter 28 should be seen primarily as providing guidelines for local authorities, which they have to adopt and adapt to suit their specific needs. The content of LA21 is not spelt out in Chapter 28: rather it is expected that each Local Action Plan will address a community's specific needs, local context and resource availability.

Box 5.3 Good governance for LA21: the ICLEI principles

- *Ecological limits:* all citizens and communities must learn to live within the earth's carrying capacity.
- *Partnerships:* alliances among all stakeholders need to be established to ensure collective responsibility, decision making and planning.
- *Accountability:* all stakeholders need to take responsibility for their decisions and actions.
- *Participation:* all major groups in society need to be directly involved in sustainable development planning.
- *Transparency:* all information is to be made easily available to the public.
- *Equity and justice:* environmentally sound, socially just and equitable economic development must work hand in hand.
- *Concern for the future:* plans and actions need to address short-term and long-term trends and consider the needs of future generations.

Source: http://www.iclei.org/LA21/LA21updt.htm, accessed 22 January 2004.

Box 5.4 Key steps in the LA21 process

- *Multi-sectoral engagement:* establishing a local stakeholder group to serve as the coordination and policy body.
- *Consultation:* consultation with community groups, NGOs, businesses, churches, government agencies, professional associations and trade unions.
- *Devising a Community Action Plan:* creating a shared vision and identifying proposals for action.
- *Participatory assessment:* assessing local social, environmental and economic needs.
- *Participatory target setting:* negotiations among key stakeholders or community partners in order to achieve the vision and goals set out in a Community Action Plan.
- *Monitoring and reporting:* establishing procedures and local sustainable development indicators to track progress and to allow participants to hold each other accountable to a Community Action Plan.

Source: adapted from ICLEI (2002).

From agenda to action: developments in LA21

The WSSD, held in Johannesburg in 2002, reviewed progress made since the Rio Earth Summit. As the action plan for the implementation of the Rio agreements, Agenda 21 became a chief focus of attention. LA21 was singled out for particular scrutiny. The Local Government Declaration to the WSSD acknowledged the need for urgent action at this level when it argued:

> We live in an increasingly interconnected, interdependent world. The local and global are intertwined. Local government cannot afford to be insular and inward looking. Fighting poverty, exclusion and environmental decay is a moral issue, but also one of self-interest. Ten years after Rio, it is time for action of all spheres of government, all partners. And local action, undertaken in solidarity, can move the world.
> (http://www.johannesburgsummit.org, accessed 3 October 2003)

The WSSD led to the launch of a new phase of LA21, Local Action 21. Local Action 21 is a new toolbox of quantifiable actions based on the LA21 process. It was designed to be the main implementation mechanism for LA21 for the next ten years. Local Action 21 commits local governments to an array of actions (Box 5.5).

Box 5.5 Action under Local Action 21

- Going beyond general sustainable development planning to address specific factors that prevent a sustainable future, such as poverty, injustice and social exclusion, insecurity and conflict. Attention will, in particular, be focused on creating sustainable communities and cities.
- Protecting the global common good through reducing the impact of cities on the worldwide depletion of resources and environmental degradation.
- Working towards the implementation of sustainable development action plans. This will involve anchoring principles, policies and practices in participatory governance and municipal sustainability management.

Source: adapted from http://www.localaction21.org/action_areas.htm, accessed 27 April 2005.

Distinguishing LA21 from general environmental activities

Despite the guidelines, principles, timetables and targets that have been laid down to structure LA21 activities, it has proved difficult to distinguish these efforts from general local environmental protection measures. This problem was encountered in the reporting that states make to the UN about their LA21 activities. States often list all and any environmental activities of local authorities as LA21 activities. This makes it hard to monitor actual compliance with the objectives of Chapter 28, to evaluate the impact of Agenda 21 overall and to identify how it has contributed to the promotion of sustainable development at the local level and, collectively, at the global level.

In order for local authorities' environmental activity to form part of the LA21 process, it has to have certain characteristics (Box 5.6). Drawing upon Agenda 21, ICLEI has also established a guide for determining whether action constitutes an LA21 process (Box 5.7).

Box 5.6 Characteristics needed for LA21 to be part of efforts to promote sustainable development

- A conscious effort to relate environmental effects to underlying economic and political pressures.
- Local issues and decisions have to be related to global impacts, not just in relation to the environment, but also in relation to the principles of sustainable development, such as justice and equity.

- A commitment to environmental policy integration.
- The involvement of the local community, including local stakeholders, business and organized labour.
- A commitment to define and work with local problems within a broader ecological and regional framework and within a longer period.
- Identification with Rio and the related processes that Rio has spawned.

Source: adapted from Lafferty and Eckerberg (1998a).

Box 5.7 Elements of a Local Agenda 21 process

- Led by, or includes substantial involvement of, a local government.
- Includes significant community participation and stakeholder involvement.
- Comprehensive, encompassing environmental, economic and social issues.
- Adheres to a long-term focus, including a plan, programme or set of actions for the local government and the rest of the community.
- Specific goals, implementation measures, monitoring and evaluation (e.g. audits, indicators, targets) are part of this long-term effort.

Source: adapted from ICLEI (2002).

Exploring LA21 practices

LA21 processes have been expanding worldwide since the Rio Earth Summit. Surveys conducted in 1997 in preparation for the Earth Summit + 5 showed that more than 1,800 local authorities had established an LA21 planning process in their locality. Five years later, on the occasion of the WSSD in 2002, surveys showed that an even larger number of local governments and their partners in 113 countries had adopted an LA21 framework. However, LA21 has failed, as yet, to develop in some regions, noticeably in the Middle East.

Of the worldwide LA21 initiatives reported to the WSSD, 61 per cent had established Local Action Plans. These plans address diverse priority issues, such as water management, unemployment, poverty, health and climate change. However, many of the plans deal with general development issues and not the promotion of sustainable development. In fact, applying the ICLEI criteria set out above, only 36 per cent of these Local Action Plans can be seen as having taken a comprehensive sustainable development approach to local planning, one that specifically incorporates economic, social and environmental needs (ICLEI 2002).

Surveys also show how the experience with the implementation of LA21 varies under different economic, social and regional conditions. To begin with, the degree of stakeholder involvement can vary significantly, both within the industrialized world (the UK and Spain providing contrasting examples) and between the industrialized countries and the Third World (Sweden and India providing contrasting cases). Actual involvement can range from simply providing input at the consultative stage to direct involvement in budgetary management.

In addition, priorities differ quite substantially, as would be expected, between the industrialized countries and the Third World. In Africa, for example, LA21 plans have typically prioritized poverty alleviation, while energy conservation is often a priority of European plans. Similarly, waste reduction has been identified as important for LA21 activity in Europe, whereas municipalities in the Third World are more concerned with community empowerment and education. However, water has been a common priority, irrespective of economic situation, with over 50 per cent of all municipalities identifying water resource management as a prime concern (ICLEI 2002).

Irrespective of the different social, economic and political contexts in which they operate, all local authorities face on-going problems integrating LA21 plans into the other areas of their policy remit. Environmental policy integration has proved particularly difficult at this level of government, an issue that is discussed later in this chapter. Furthermore, municipalities, even in the richer, Northern countries, often lack the resources to support LA21 activities, especially financial resources stemming from central government. Most municipalities also identified the lack of a national tax structure that rewards sustainable development practices as a key obstacle (ICLEI 2002).

In the next few pages the works of Eckerberg and Lafferty and of the ICLEI are used to present some case studies. These are grouped according to region, beginning with Western Europe.

Implementing LA21 in Western Europe

Surveys show that the vast majority of LA21 activity is to be found in Europe. Over 5,292 European municipalities have become involved, accounting for over 80 per cent of the worldwide engagement. However, despite the relatively advanced state of LA21 work in Western Europe, there are still differences in the timing and the extent of LA21 activities between different European countries. Some countries began their LA21 activities relatively soon after the Rio Summit and many, if not all, of their local authorities went on to engage in LA21 initiatives. In contrast, other countries began late and only a few of their local authorities have since become engaged.

Figure 5.1 *Making links: Cardiff County Council, Wales, promotes sustainable development as a shared responsibility, as seen in a leaflet from its Sustainable Development Unit*

Courtesy: Cardiff Council

Box 5.8 Engagement with LA21 in Western Europe

- *Early and many:* Sweden, the UK, the Netherlands.
- *Later and many:* Denmark, Norway, Finland.
- *Later and few:* Greece, Austria.
- *Latest and least:* Spain, Italy, Ireland, France.

Source: Eckerberg and Lafferty (1998).

There are several reasons for these differences. First, a close correlation exists between LA21 activity and the type of local government system in a country (Lafferty and Eckerberg 1998a). In the northern European system (Norway, Sweden, Denmark, Finland), local authorities have a high degree of autonomy, are able to get their own funds through raising taxes and have a broad scope of power in relation to environmental matters. With their long-standing tradition of local government autonomy, Nordic countries were front runners in LA21. The Netherlands was able to match this engagement, despite having a more centralized tradition, due to its highly developed and participatory approach to environmental planning. The Middle European system (Germany and Austria), which is a federal system where local authorities are small and have varied powers, has experienced a slow start to LA21 activity. In the Anglo-Irish system there are relatively large local authorities but they have few powers and weak financial bases and are dependent on central government. This system, as well as the Napoleonic system (France, Spain, Italy), where there is a high degree of central government control, have both resulted in late and few LA21 initiatives. The UK proved the exception, as LA21 provided an opportunity for local government to overcome some of the centralizing tendencies of the reforms introduced during the years of Conservative government under the leadership of Margaret Thatcher.

Second, to be successful, local initiatives need the support of national policies and national funding. Other factors that influenced both the timing and the extent of LA21 activities included the presence of an active and politically mobilized population, including environment NGOs, interested and motivated civil servants and local politicians, and involvement with international networks. The support of local political and administrative decision makers was found to be particularly crucial. Where participatory practices are poorly developed, as in Spain, late and weak responses to LA21 are typical (Eckerberg and Lafferty 1998). As will be seen in the case studies below, the absence of civil society organization also proved a barrier to LA21 in India. LA21 processes also depend on being built into existing organizations that are themselves concerned with sustainable development issues.

In Western Europe the work of Agenda 21 relies heavily upon the 1998 Convention on Access to Information, Public Participation in Decision-making and Access to Justice in Environmental Matters, known as the Aarhus Convention. This elaborates Principle 10 of the Rio Declaration, which stresses the need for citizens' participation and for access to information on the environment held by public authorities. Drawing upon this principle, the Aarhus Convention seeks to promote sustainable development through granting procedural rights. Such rights include citizen access to information, the right to public participation and access to justice in environmental matters. It is premised on the belief that granting procedural rights will enable citizens to participate directly in environmental decision making, thereby enhancing the quality of environmental policy. The convention also focuses on the international governance issues, in particular tackling the democratic deficit in the negotiation of international treaties and in the operations of international institutions such as the World Bank and the WTO. Both these institutions are infamous for their secrecy and distance from the public (Bell 2004), issues discussed in Chapter 7.

The Aarhus Convention has been signed by over forty European countries and by the EU and its member states. The procedural rights granted under the Aarhus Convention represent a genuine step in enhancing participation and in democratizing public policy making. In this sense, the convention represents an example of the new governance approach to the promotion of sustainable development. Nevertheless, the convention has a rather limited notion of participation, stressing the primacy of representative institutions and giving a constrained role to public participation. It also places strong emphasis on local decisions, rather than on regional, national or international decisions, that may operate at a strategic level (Bell 2004). Thus the convention is subject to both the strengths and the limitations of the use of participatory democracy for the promotion of sustainable development.

Case study: Germany

A study of German local authorities found three different types of LA21 activity (Box 5.9; Moser 2001). Despite the growth of LA21 activity, traditional interests and lobbying still retain more influence on public policy at the local level than do new, mobilized environmental interests. In contrast to the somewhat depressing German case, Sweden has proved to be a leader in LA21 activities.

Box 5.9 LA21 responses by German local authorities

- *Type 1*: clear emphasis on practical demonstration projects in selected areas, such as establishing regional marketing and promoting solar energy.
- *Type 2*: this approach favoured the integration of the urban development programmes of LA21 into an open concept of future development.
- *Type 3*: this involved a more programmatic approach, along the lines of modern strategic planning.

Source: adapted from Moser (2001).

Case study: LA21 in Sweden

Sweden is regarded as an LA21 leader state (Eckerberg and Forsberg 1998; Eckerberg 1999; Rowe and Fudge 2003). Almost 100 per cent of municipalities are engaged in LA21. National ideology, especially a strong commitment to the common good, advanced environmental policy frameworks and the availability of a wide range of financial tools have all helped promote LA21 work in Sweden. This has helped to create a more enlightened view of environmental and development issues, encourage more cross-sectoral cooperation and foster networking within and between municipalities (Rowe and Fudge 2003).

Box 5.10 LA21 in Sweden

Background

- Traditional commitment to high levels of public ownership of land, utilities and resources.
- Strong environmental protection measures, linked with the economy and technological innovation.
- High level of social welfare spending, dependent on high taxation.
- Strong devolved government.
- Relative societal homogeneity, deeply rooted in traditions of equity and cooperation, and a clear and shared sense of the public good.

Features of environmental protection policy: leader environmental state

- High-quality environmental research.
- Environmental monitoring linked with established indicators.
- Strong environmental legislation and frameworks for administration.

- Inclusion of environmental considerations in physical planning.
- Commitments to the 'polluter pays' principle.
- Support and fiscal mechanisms for linking environmental policy and practices.

LA21 work

- Early adoption of LA21 in the 1980s.
- Appointment of LA21 officers at the municipal level.
- Central government financial assistance.
- National Agenda 21 Forum established.
- Public knowledge about and interest in LA21 widespread.

Problems

- Only 3 per cent public participation in an LA21 project.
- Lack of, conflicting or unclear goals, visions, tools and guidance from central government.
- Strongly sectoral policy making at both national and municipal level remains unchanged, in spite of LA21.
- Funding for LA21 varies widely across municipalities.
- The Swedish social democratic model is under threat from globalization, modernization and cultural changes; recent immigration has increased social diversity but also heightened tensions in Swedish society.

Sources: adapted from Eckerberg and Forsberg (1998); Eckerberg (1999); Rowe and Fudge (2003).

The urban dimension: planning for sustainable development

The Brundtland Report stressed the need to focus on cities as critical locations for the promotion of sustainable development, especially given that most of the world's future population will live in urban areas. The development of cities has resulted in a dramatic shift in the spatial and material relations that we have with the rest of the ecosphere. Most of the human population now live and work far from the land and the biophysical processes that support them (Rees 1999). Cities are also a drain on the environment, as they require ever greater quantities of food, material commodities and energy to sustain their inhabitants. The concept of ecological footprint is used to capture the extent of this drain. As cities grow, city authorities, in both the industrialized countries and the Third World, need to find ways to contain the ever growing burden that cities place on the immediate environment and, ultimately, on the global commons (Satterthwaite 1999). The focus on cities has also led to a distinction between 'Green' and 'Brown' environmental agendas. The Green agenda focuses on reducing the impact of urbanization on the natural ecosystem, while the Brown agenda looks at the need to address environmental threats to health from overcrowding, lack of sewage treatment, inadequate waste disposal and water pollution. The Brown agenda is reflected in the prioritization of WEHAB initiatives at the Johannesburg Summit.

However, the very factors that make cities weigh so heavily on the ecosphere – the concentration of population and consumption – give them enormous economic and technical advantage in the quest for sustainable development (Rees 1999). Cities have become important sites in the quest for ways to promote sustainable development, drawing in issues of spatial planning, housing design, transport and land-use planning. There is also new emphasis on the advantages that cities can offer by virtue of their economies of scale. These lower the cost *per capita* of providing infrastructure, increase the range of options for material recycling and reuse, reduce the *per capita* demand for occupied land, allow greater possibilities of electricity co-generation and offer greater potential for reducing fossil fuel consumption through the provision of public transport (Rees 1999). The challenge is for both the urban ecology movement and for urban planners to find ways in which these advantages can be realized. LA21 represents a prime example of UNCED efforts to meet that challenge.

There have been two UN 'Habitat' conferences exploring the urban dimension of sustainable development. The EU has also taken action in this area. The adoption of the 1998 *Sustainable Urban Development in the European Union: A Framework for Action* (CEC 1998a) and the communication *Towards a Thematic Strategy on the Urban Environment* (CEC 2004a) have both helped promote a policy agenda on urban sustainability within the EU. The 2004 strategy is particularly important and it aims to promote LA21 by addressing a wide range of urban issues, such as noise, poor air quality, heavy traffic, neglect of the built environment and lack of strategic planning. The overall aim of the strategy is 'to improve the environmental performance and quality of urban areas and to secure a healthy living environment for Europe's urban citizens, reinforcing the environmental contribution to sustainable urban development while taking into account the related economic and social issues' (CEC 2004a: 4).

Action has also developed at the national level. In the UK the promotion of Sustainable Cities has entered into mainstream planning, regeneration and development agendas. A number of environmental and urban policy objectives have been developed to help the implementation of this new approach to urban development (Couch and Dennemann 2000; Bulkeley and Betsill 2005). Under LA21 processes the sub-national level also has a key role to play. Local authorities are increasingly charged with the task of promoting sustainable development in urban settings, especially by drawing transport and land-use planning into new, environmentally sensitive urban area plans (Smith *et al.* 1998; Kenny and Meadowcroft 1999).

The European Sustainable Cities and Towns Campaign is a good example of UN-sponsored action to address the urban dimension of sustainable development. It involves both sub-national engagement and participation in a parallel process of networking across municipalities. Much of this networking is aimed at identifying exemplars of 'best practice', from which lessons can be learned and policies transferred from one city to another. In this sense, it also represents an example of the new governance approach to policy making and implementation. Framing the promotion of sustainable development within a discrete spatial scale, while participating in transnational municipal networks engaged in promoting urban sustainability, has been dubbed 'new localism' (Marvin and Guy 1997).

Box 5.11 The European Sustainable Cities and Towns Campaign

The European Sustainable Cities and Towns Campaign was launched at the end of the first European Conference on Sustainable Cities and Towns, held in Ålborg (Denmark) in May 1994. The ICLEI was one of the principal organizers of the conference. The founding document of the campaign is the Ålborg Charter, which outlines what is understood by local sustainable development and contains a commitment to engage in LA21 processes. More than 1,000 local authorities participate in the campaign.

The Ålborg Charter has the following goals:

- protection of the countryside and building up resources;
- social equity;
- sustainable development;
- sustainable mobility;
- policy measures to reduce greenhouse gas emissions;
- prevention of toxic emissions;
- local autonomy;
- popular participation.

In addition, the work of the campaign is guided by the 1996 Lisbon Action Plan, which outlines twelve action points for preparing local government for LA21.

At the third European Conference on Sustainable Cities and Towns in Hanover in February 2000, 250 European mayors issued the Hanover Call. This declares local sustainable development as their particular responsibility and an issue of highest political priority.

continued

Five international networks and associations of local authorities support the European Sustainable Cities and Towns Campaign: the Council of European Municipalities and Regions; the United Towns Organization (now the World Federation of United Cities); Eurocities; the Healthy Cities Project of the World Health Organization; and ICLEI.

These five networks undertake a variety of activities to support the campaign. Activities include awareness raising and consultation, holding good practice seminars and providing technical guidance and training. During 1998/99, for example, ICLEI organized regional conferences in Turku, Sofia, Seville and The Hague, addressing specific issues in northern, eastern, southern and western Europe, respectively.

Source: adapted from http://www.iclei.org/europe/la21/sustainable-cities.htm, accessed 24 January 2004.

The Sustainable Cities agenda and the promotion of new localism have gained significant ground in the UK. While supported by the Prime Minister's Office, much of this is due to the enthusiasm with which local authorities adopted LA21. In the UK LA21 was seen as a way in which local authorities could reclaim the policy ground lost during the Conservative governments of the 1980s (Bulkeley and Betsill 2005). There is also a strong emphasis in contemporary Dutch planning on the local level and on improving the 'livability' of urban areas (Eckerberg *et al.* 2005). There are parallels between this development and the focus on sustainable communities that emerged in the US in the 1990s. This emphasized the efficient use of urban space, minimizing the consumption of essential natural capital, multiplying social capital and mobilizing local government and citizens to meet these ends (Roseland 2000).

LA21 in the Americas

The United States

The work of the President's Council on Sustainable Development can be seen as the US response to the Earth Summit and thus as the country's national Agenda 21 activity. President Clinton established the council in 1993. Its remit was to:

Advise the President on the next steps in building the new environmental management system of the 21st century . . . further developing a vision of innovative environmental management that fosters sustainable development (environment, economy and equity), and recommending policy improvements and additional opportunities to advance sustainable development.

(http://clinton2.nara.gov/PCSD, accessed 27 April 2005)

The Council's report, *Sustainable America: A New Consensus*, however, did not refer to LA21 by name (PCSD 1996). Nevertheless, the council echoed the Earth Summit's support of local government intervention to promote sustainable development. It also supported action on a range of fronts, including in relation to developing community-driven strategic planning and collaborative regional planning, improving community and building design and decreasing urban sprawl (Bryner 2000).

The President's Council also recognized the importance of local community action. However, in the US, there is no coordinated or explicit campaign to encourage local governments to undertake comprehensive, multi-issue planning and implementation processes analogous to the LA21 work that has begun in other countries. Nevertheless, several US local governments have undertaken processes somewhat similar to LA21.

Over the past few years, many governmental and community actions and programmes in the US have invoked terms like 'sustainability' and 'sustainable development'. For example, research by Public Technology Inc. has uncovered almost 1,500 initiatives in more than 700 US cities and counties that are labelled 'sustainable' (http://www.iclei.org, accessed 22 January 2004). However, a great many of the projects deal only with a single issue, such as economic revitalization, transport alternatives, energy conservation, a specific environmental concern or the redevelopment of a particular site or district. These programmes are not comprehensive, in that they do not include environmental protection or economic, social and community development as integral parts of their plans. In addition, many do not involve local government agencies. As such, it is difficult to consider that these initiatives, by themselves, constitute an LA21 process.

The ICLEI applied the criteria set out in Box 5.7 in an effort to determine whether the US local governments were undertaking work analogous to LA21. They were able to identify only twenty-two US municipalities making efforts similar to an LA21 programme. These were the only municipalities engaged in a comprehensive, long-term effort to promote sustainable development by: integrating planning and action in the environmental, economic and community spheres; being led or supported by local government; including significant community and stakeholder involvement; and specifying a set of activities to achieve its goal. However, the majority of these twenty-two municipalities are only in the initial stages of implementation of LA21-type activities, and the most advanced municipalities do not necessarily use the term 'sustainable development'.

However, it has been discovered that local government has become engaged in LA21-type processes in the US through indirect routes. The indirect route is where the local government's planning mission and day-to-day operations have incrementally evolved to include all the attributes of sustainable development comprised in LA21. Although it has not adopted a plan or policy statement that is labelled 'sustainable development', its operations have been modified over several years to the point where they integrate sustainable development objectives (http://www.iclei.org, accessed 22 January 2004).

Latin America

There are over 119 LA21 initiatives scattered across seventeen countries in Latin America. The bulk of them are in Brazil, with Chile, Ecuador and Peru providing the second largest numbers. Most of these initiatives have occurred without campaigns at the national level. The most common priority in Latin America is community development. Latin America is the only region to identify tourism as an LA21 priority and to include heritage and cultural preservation as areas of activity. However, while the region has a very high rate of stakeholder involvement, ethnic minorities, indigenous peoples, trade unions and youth are underrepresented or absent from these stakeholder groups (ICLEI 2002).

Case study: Quito, Ecuador

Quito is the second largest metropolitan area in Ecuador. The local government is responsible for urban development planning, drinking water, sewage and other environmental services, the construction and maintenance of local roads and public spaces. The city has numerous environmental problems, including air pollution and water contamination. In 1990 the municipality created the Department of Environmental Quality Control, and the Metropolitan District Law was passed in November 1993, allowing environmental problems to be addressed through municipal legislation. The law divided the municipality into North, Centre and South administrative zones. The South Zone is the poorest part of the city, with many illegal settlements, the most rapid population growth in Quito, the lowest service provision and the highest concentration of industry.

Following the new law, the South Zone administration organized meetings with churches, NGOs, *barrio*-level committees, the armed forces, industries, municipal departments and municipal service companies. This was the first ever community consultation process held in the city. In addition, public forums were held to help reach consensus on priorities for action. One of the products of this consensus-building process was the South Zone Integral Development Plan. After considering four projects for LA21 activity within the plan, the South Zone Ravine Restoration Project was chosen as the most urgent, and was launched in January 1996 (Box 5.12).

Box 5.12 The South Zone Ravine Restoration Project, Quito, Ecuador

The South Zone Ravine Restoration Project was the first urban planning exercise in Ecuador that operated on the premise of community consultation and participation. The object of the project was to involve stakeholders in the development and implementation of a plan of action that would restore the ecological balance of the ravines, and improve the quality of life of residents in the surrounding *barrios*. The project was linked with the municipal statutory planning process through the South Zone Integral Development Plan.

A Coordination Committee was first established, which acted as the stakeholder group. Members included local government, churches, residents' associations, NGOs and academia. Despite receiving invitations, private-sector industries did not participate.

The main functions of the Coordination Committee were to set the agenda, evaluate projects, draft proposals and define strategies. Later, the committee coordinated the ravine project and dealt with its financial management.

A document entitled 'A Vision for the Future of the South Zone' was then drawn up. It formed the basis of the South Zone administration's strategic plan. In addition, an Environmental Assessment Report on the ravines was produced in 1996. This identified the priority issues of residents living around the ravines and the most important risks they faced. The risks posed by industrial waste were identified as the main problem, with garbage and the disposal of untreated waste from a local abattoir as the second and third most important risks. The assessment recommended action to limit pollution from source and to establish closer collaboration with the Department of Environmental Protection.

Various partners committed funds to the South Zone Ravine Restoration Project. An Action Plan was set up, focusing on long-term solutions. First, the plan structured the integrated management of the ravines through the creation of a legal, organizational and financial framework. Second, an education and information programme was created. The plan also called on industries to eradicate ravine pollution at source. These actions are on-going.

Source: adapted from http://www.iclei.org, accessed 22 January 2004.

LA21 in the Asia-Pacific region

There were 674 LA21 initiatives across seventeen countries in the Asia-Pacific region reported to the WSSD. The presence of a national campaign proved particularly important in stimulating activity in Australia, Japan and South Korea. One of the noticeable features of LA21 in this region is the strong emphasis on initiatives dealing with environmental protection. Within that, the most common priority was natural resource management, followed by air quality, water resources and energy management. However, while stakeholder involvement is generally high, youth involvement and participation of ethnic minorities has been very low in this region.

In Third World countries, stakeholder groups share in LA21 decision making to a much greater degree than they do in the industrialized world. However, the experiences they have, the context within which they work and the priorities they set differ in fundamental ways from those encountered in LA21 activities in the industrialized world.

Box 5.13 Pimpri Chinchwad, India

Pimpri Chinchwad is one of the largest industrial cities in India. It has over 2,000 engineering, chemical, rubber, pharmaceutical and automobile factories. Of its population of 517,300, approximately 100,000 residents live in slum settlements, which are beset with environmental and health problems. Most of the city has an open drainage system, and raw sewage and industrial effluent are dumped into the Pawana river.

A large portion of the factory work force resides in neighbouring villages and the city of Pune, while the population of Pimpri Chinchwad is composed mainly of new migrants. The resulting lack of cultural identity in the city is reflected in the alienation of citizens from the process of community development planning. The local community has little involvement in the running of the city beyond electing its officials. The industrial sector has not been involved in city development planning.

Creating community awareness was the first step in establishing an LA21 process for the city. This began in January 1995 with a vigorous publicity campaign that involved newspaper advertisements, radio broadcasts and pamphlet distribution.

The next step, establishing a dialogue between the city and the citizens, was undertaken by organizing two public meetings. About 150 citizens, fifteen to twenty councillors and ten to fifteen officials attended each. These meetings were the first of their kind to be held in the city. However, citizens were confused about the purpose of the meetings and their role in the planning process. Some expected to receive financial aid or to have their personal problems addressed. At the same time, many educated and more affluent citizens believed the LA21 process was meaningless talk about environmentalism.

Creating community partnerships through the establishment of a stakeholder group was the next step. The stakeholder group consisted of twenty-five people, including the mayor, the deputy mayor, the opposition party leader, the standing committee chair, the ruling party leader and people drawn from medicine, academia, the media, the natural sciences and voluntary organizations.

Briefing workshops, a visioning exercise and brainstorming sessions were held to facilitate education and teamwork within the stakeholder group. The stakeholder group reviewed the findings of the community consultation process. However, lack of experience in forms of participatory democracy resulted in problems with the community consultation process. Conventional methods of communication with the community were also found to be inadequate. Consequently the city commissioned a private consultancy service to facilitate the process. Between 1996 and 1997 they used field observations, interviews, group meetings, focus group discussions and structured questionnaires. This community consultation process was the first of its kind in the city, involved more than 13,000 citizens and reached all sections of society, including the poor, lower castes and women.

The priorities and issues identified by the community consultation process were reflective of a citizenry that was struggling to survive and unable to meet basic needs such as drinking water, food, shelter and health services. The two main issues to emerge were waste management and the improvement of slum areas. Slum improvement includes addressing bad housing development and poor municipal services (including drinking water and drainage). This helped to set priorities for what is involved in the promotion of sustainable development in the city: providing immediate relief to the neediest, while also adopting a long-term perspective on planning.

Source: adapted from http://www.iclei.org/LA21, accessed 24 January 2004.

New Zealand

The planning process in New Zealand, and hence the context within which LA21 operates, underwent major reforms that were started in 1984 and continued with another round of local government reform in 1989. These reforms resulted in a massive reduction in the role of central government. They were driven by neo-liberal, market-based economic principles. The reforms had a noticeable social impact, especially following the significant reductions in welfare benefits in the early 1990s. From the mid-1980s until the early 1990s the system of environmental management was also radically changed, including at the legislative and administrative levels (Bührs 2003).

Box 5.14 LA21 in Hamilton, New Zealand

The city of Hamilton is the fifth largest urban centre in New Zealand. The quality of Hamilton's environment is relatively high. Air pollution is minimal; efforts are being made to reduce, reuse and recycle waste; issues such as contaminated sites and liquid waste disposal into river and groundwater systems are being addressed.

Hamilton City Council formally adopted the twin principles of responsiveness to the community and responsibility to meet community needs as part of a major administrative reform undertaken in 1987. Later, the Local Government Reform Act 1989 required local government to recognize the existence of different communities and their identities, values and rights, and to ensure their effective participation in local government.

In 1993 Hamilton formally adopted the principles and objectives of Agenda 21 and the council took its first steps to develop an LA21 strategic plan and planning process. The objective was to produce a twenty-year plan for the city, using the principles and philosophies of Agenda 21. The plan was to become known as Hamilton's Model Communities Programme.

In November 1994 workshops to identify issues and create long-term visions were held with three groups of community partners: planning partners, representatives of governmental departments and agencies from Hamilton and the Waikato region and more than 230 key community organizations. All participants were invited back to a consensus forum in December 1994, where information from the initial workshops was presented. The output of this forum was summarized in sixteen Visions for Hamilton, subsequently known as 'The Cloud', because of their graphic presentation.

Five task forces were created in May 1995 to look at specific development issues. The five task forces met weekly to develop various options or scenarios for Hamilton's future development. However, as the process continued it was difficult to keep the sixteen visions in the forefront, and it was decided to focus on five major areas: the environment, city growth, community development, economic development and the central business district.

The output from the task forces was made available for public comment in June and July 1995, through a travelling road show, a mail-return questionnaire in a special edition of the council's newspaper delivered to all households, telephone surveys of 500 households, presentations to groups and organizations, news bulletins and interviews with the media, youth meetings and a telephone hot line.

Between December 1994 and June 1995 a separate consultation process was held with Maori people. The key results of this parallel consultation process were merged into the draft Strategic Plan. Maori were involved in the general process as well.

The Strategic Plan was developed. Following its public launch in November 1996, the principal planning partners signed a formal agreement confirming their commitment to work with the council to further the principles and objectives of LA21 and the visions set out in the Strategic Plan. The partnership group meets at least once a year.

Source: adapted from http://www.iclei.org/LA21, accessed 21 January 2004.

Promoting LA21 in least developed countries: LA21 in Africa

Local governments in Africa face challenges that are often well beyond their management capacity, including conflicting interests between groups, often resulting in violent civil unrest. Demographic pressures, the Aids crisis, inadequate infrastructure and very limited resources for service delivery and planning add to these difficulties. In many countries, urban environmental management has been added to a long list of municipal responsibilities. The inability to manage urban development effectively is rapidly increasing the human risks associated with poor housing conditions, uncollected solid waste and over-consumption of fresh water, untreated waste water and urban air pollution.

Despite the difficult conditions under which local authorities in Africa work, LA21 processes have been reported in 151 municipalities in 28 African countries. However, only South Africa has established a national campaign to facilitate LA21 activities. Economic development remains an overriding concern of LA21 activities, with a specific focus on poverty alleviation. In fact 90 per cent of municipalities have prioritized poverty alleviation as the main objective of their LA21 initiatives. Stakeholder groups remain weak: women have participated in less than half the processes. and ethnic minorities have participated in only 3 per cent.

Box 5.15 African Sustainable Cities Network

The African Sustainable Cities Network was established in 1995 within the framework of LA21. It was established by a group of elected officials and senior professional staff from African cities. The network helps capacity building and exchanges between the participating cities. It provides support to help develop locally appropriate, sustainable responses to local environmental and social problems.

---

continued

The network is coordinated by ICLEI, which sees it as a tool to enhance the capacity of local authorities in Africa to institute participatory environmental planning. ICLEI support can be divided into two phases:

- *Phase 1 (1995–6), Assessment*: ICLEI undertook an environmental management needs assessment; supported a small group of five African local governments to develop their LA21 planning efforts; facilitated information exchange.
- *Phase 2 (1997–2000), Implementation*: the objectives of the second phase are to: educate public administrators, elected officials and local citizens about the technical aspects of key environmental issues that are affecting the quality of life in their local communities; help the implementation of new national environmental policies that are delegating specific environmental responsibilities to local authorities; support the creation of LA21 initiatives.

Examples of activities undertaken through the network:

- *Information exchange network*: a Web site was set up to promote and facilitate information sharing among core network cities.
- *Pilot project component*: seven core cities were provided with training, programme support and small grants for local pilot projects.
- *North–South partnership relations*: this involved establishing partnerships with municipalities in Europe.
- *Publications*: for example, a *Guide to Environmental Management for Local Authorities in Africa* was published in 2000.
- *Regional conferences*: these were Zimbabwe (1998) and Kenya (1999), to bring together project participants, to share information, to stimulate discussion and to debate, build capacity and further develop the role of the network.
- *Development of performance indicators*: this involves defining indicators to measure progress in the development and implementation of Local Action Plans. To assist in this process, ICLEI developed *A Guide to Results Indicators for Local Sustainable Development Planning* (1998) and provides training on its use.

The second phase of the African Sustainable Cities Network proved to be a catalyst for the development of LA21 planning and regional networking.

Source: adapted from http://www.iclei.org/LA21, accessed 23 January 2004.

Limitations of LA21

At one level, LA21 can be seen as primarily concerned with procedural issues, with putting methods in place in order to promote sustainable development at the local level. In particular, it targets planning and aims at ensuring the integration of sustainable development into local authority planning processes. However, local authorities do not act in isolation and their capacity to shape policy and its outcomes is shaped by a number of factors. To begin with, account has to be taken of the various levels of governance (local, national and international) through which economic, social and political processes interact to shape the prospects for the promotion of sustainable development at the local level. The impact of LA21 has to be viewed in this wider governance context. This means taking account of the distribution of power and authority between central government and sub-national authorities. National policy frameworks, for example, can have a major influence on the local planning process, as do statutory obligations, statutory municipal development plans and national budget priorities. National taxation policy can also act as a barrier, especially when there are subsidies and other tax incentives that encourage unsustainable practices, particularly with respect to resource use. In many countries, central government maintains control of local budgets, which makes it difficult to coordinate national investment plans and local LA21 priorities. The ability of local government to generate revenue is also regulated and restricted by national policies. These factors all combine to shape the capacity of local authorities to implement LA21 initiatives successfully (Gilbert *et al.* 1996).

While effective LA21 initiatives require local authorities to gain control over their development, globalization is accelerating investment and development activities by external actors, such as transnational corporations. Often these have only a minimal incentive to be accountable and committed to local development strategies (Gilbert *et al.* 1996). Thus the ability of local authorities to structure their development in a sustainable way is limited, not just by the entanglement of local governance structures in the national political and economic system, but also by the system operating at the international level. Local authorities in industrialized countries, for example, struggle to find ways to deal with waste generated by consumer products and packaging. While this accounts for a large portion of the local solid waste stream, local governments have little direct control over the products that are sold and used in their jurisdictions. In contrast, in the Third World, basic life needs, for health, sanitation and water, structure the priorities of LA21 activities, yet the very international conditions that make these a problem are also barriers to their effective resolution. The existing global structures of political and economic power stand in direct opposition to the construction of local development models based around LA21 principles. As such, adapting economic, ecological and social interests to the concerns of sustainable development is possible in only a limited fashion at a local level. When account is taken of the fact that a great deal of power and authority lies at the global level, the scope for local authority action becomes limited.

The focus on the urban dimension of sustainable development provides a particularly good example of the dangers of divorcing the urban from other levels through which environmental governance is mediated. The rise of 'new localism' in particular implies the danger that the local can become a 'black box' disconnected from the global, international and national contexts with which localities are framed (Marvin and Guy 1997; Bulkeley and Betstill 2005).

Conclusion

LA21 is both a procedural quest (in relation to planning) and a highly political process. LA21 initiatives are designed to open up the planning process through facilitating dialogue and information exchange between groups drawn from across society, the economy and the political sectors, including from within public administration. The advantage of this exchange is that it can increase respect for differing points of view, while at the same time it helps participants to develop a more complex understanding of the issues involved in seeking to reconcile local environmental, social and economic needs. In addition, the extensive public involvement that is an integral part of the LA21 process helps develop a sense of community purpose, giving the community new confidence in its ability to shape its future. Engagement in LA21 can also help create a trained and educated forum for environmental and social issues, which can be of wider importance within a locality. In other words, within an LA21 process, participation has a functional logic: participation can enhance the quality of decision making, leading to more alternative options, more systematic identification of problems and a wider range of suggested solutions. Implementation of sustainable development plans and their acceptance, or legitimacy, at the local level is easier to achieve with expanded participation.

However, LA21 needs to be seen as more than this. The requirement that LA21 should enhance the participation of civil society, especially of women and youth, means that LA21 is as much about democratic reform as about planning procedures. As such, it has a strong normative aspect. Nevertheless, while participation can enhance legitimacy, lead to more innovative solutions and encourage more successful implementation, it none the less raises the thorny issues of the ability of normal citizens to take effective part in decision making, the capacity of planning processes to function effectively under conditions of participation, and the legitimacy of expanded stakeholder democracy. There is also the fact that LA21 activity is limited by the wider economic, political and institutional processes that operate at the national and international levels to constrain the ability of the local level to promote sustainable development in isolation.

Summary points

- LA21 can be distinguished from general environmental policy, because LA21 plans are meant to be built upon a *new* understanding of the relationship between the environment and development, built around the concept of sustainable development.
- By focusing on LA21, attention is drawn to the local scale and to the need for action that is responsive to the specific needs of a local place and its community.
- LA21 links planning, including in relation to land use, urban design and transport, and the promotion of sustainable development, particularly in urban settings and in the construction of the cities of the future.
- LA21 also helps to focus on the level of social organization where the consequences of environmental degradation are most keenly felt and where successful intervention can make an immediate difference to the quality of life.
- The UNCED process has given local authorities, through LA21, a pivotal role in promoting sustainable development.
- Case studies show that LA21 activities have been growing throughout the world, but most noticeably in Europe. Priorities differ, with African countries more concerned about poverty, while industrialized countries are preoccupied with energy and waste. Stakeholder involvement also differs. In northern Europe, local authorities can draw upon highly developed civil society networks, familiar with local political and policy processes. In contrast, developing countries lack such highly developed civil society structures. Consequently, LA21 processes had to concentrate on education and information, and consultation is a slower and a much more deliberate process. Capacity building remains a major task, as seen, for example, in the African Sustainable Cities Network.
- There are both instrumental and normative arguments for the participatory practices that form the core component of LA21 practices. In this sense, LA21 is about establishing good governance practices for the promotion of sustainable development.
- LA21 builds on the argument for direct democracy and emphasizes direct involvement in substantive decision making on the part of the wider public. LA21 is also premised upon the principle of shared responsibility, which, in turn, requires a redefinition of the relationship between government, including at the local level, civil society and the economy. However, LA21 raises the thorny issue of the democratic nature of participatory practices.

- Multi-level governance processes and structures shape the capacity of local actors to act. There are thus limits to the role that local authorities can play in promoting sustainable development isolated from action across all levels of governance.

Further reading

Summary of LA21 activity

Dodds, F. (1997) *The Way Forward: Beyond Agenda 21*, London: Earthscan.

International Council for Local Environmental Initiatives (2002) *Local Governments' Response to Agenda 21: Summary Report of Local Agenda 21 Survey with Regional Focus*, Toronto: ICLEI.

Osborn, D. and Bigg, T. (1998) *Earth Summit II: Outcomes and Analysis*, London: Earthscan.

Studies of LA21 activities

Hens, L. and Nath, B. (2003) 'The Johannesburg Conference', *Environment, Development and Sustainability*, 5: 7–39.

Lafferty, W. (ed.) (2001) *Sustainable Communities in Europe*, London: Earthscan.

Lafferty, W. and Eckerberg, K. (eds) (1998) *From the Earth Summit to Local Agenda 21: Working towards Sustainable Development*, London: Earthscan.

Analysis of the structural barriers to LA21 activity

Girardet, H. (ed.) (1996) *Making Cities Work: The Role of Local Authorities in the Urban Environment*, London: Earthscan.

Planning, space and sustainable development

Kenny, M. and Meadowcroft, J. (eds) (1999) *Planning Sustainability*, London: Routledge.

Web sites

http://www.iclei.org/LA21/ for information on LA21.
http://www.un.org/events/wssd WSSD portal.

Part III

The promotion of sustainable development in different social, political and economic contexts

6 High-consumption societies

The responsibilities of the European Union

Key issues

- Legal and declaratory commitment.
- Environmental Action Programmes.
- Ecological modernization, eco-efficiency and decoupling.
- World Business Council for Sustainable Development.
- EU understanding of sustainable development.
- Environmental policy integration.

This chapter explores the promotion of sustainable development in the EU. It begins by looking at the EU's strong declaratory commitment and then pays attention to the policy and strategy documents that frame the EU's efforts. What is meant by ecological modernization is then explored, and why it has taken hold within the EU and its relation to sustainable development. The distinctive features of the EU's understanding of sustainable development are outlined, and how that understanding differs from the more fundamental changes envisaged by the Brundtland Report. Examination of efforts to meet the challenge of environmental policy integration at the sectoral level provides insight into the difficulties surrounding implementation efforts. While implementation may be weak, there is still need to understand the significance of the EU's commitment.

Legal obligation

European Community initiatives in the field of environmental protection began after 1972. At the time, member states were influenced by the growing international concern about the environment. They declared that economic expansion was not an end in itself, but should result in an improvement in the quality of life as well as the standard of living (Baker 2000). In 1986 the Single European Act formally recognized the EU's role in environmental protection. By this time, the Commission of the European Communities (the Commission), the main body initiating policy in the EU, was keen to develop environmental policy at the EU level. It feared that the strengthening of environmental legislation by member states, in response to increasing domestic mobilization around environmental issues, would act as a barrier to European free trade.

Gradually there was a shift of policy focus from general environmental protection measures to the promotion of sustainable development. Treaty modifications, including the Maastricht Treaty (1992), reflected this. The Amsterdam Treaty (1997) called for 'balanced and sustainable development of economic activities', and it adopted environmental policy integration as a key means of achieving sustainable development. More important, the Treaty of Amsterdam made sustainable development one of the *objectives* of the Community, along with economic and social progress. This makes it applicable to the general activities of the EU, not just its activities in the sphere of the environment. Sustainable development has the status of a guiding principle of the European integration process. The Treaty of Nice (2000) confirmed this. The draft treaty establishing a constitution for Europe also views sustainable development as a principle having general application (European Community 2004).

Because of these modifications, through the Single European Act (1986) and the Maastricht (1992), Amsterdam (1997) and Nice treaties (2000), there is probably no single government or other association of states with such a strong 'constitutional' commitment to sustainable development as the EU. Sustainable development is now a norm of EU politics, both domestically and internationally (Baker and McCormick 2004).

Environmental action programmes

EU environmental policy is framed by medium-term Environmental Action Programmes (EAPs), of which there have been six to date. They are drawn up by the Commission and provide the strategic focus needed to turn declaratory and legal commitments into actual policy. The Fifth EAP engages explicitly with the task of promoting sustainable development, although the understanding of what this entails is strongly influenced by the previous four EAPs.

The First EAP (1973–6) acknowledged that economic growth was not an end in itself (CEC 1973), while the Second EAP (1977–81) referred to the limits to growth stemming from natural resource limitations (CEC 1977). It stressed that neither economic development nor the 'balanced' expansion of the Community could be achieved without environmental protection, and it affirmed that 'economic growth should not be viewed solely in its quantitative aspects' (CEC 1977: 8). The Third EAP (1982–6) stressed the links between environmental policy and the Community's industrial strategy, arguing that environmental protection measures could stimulate technological innovation (CEC 1983). This argument was to prove decisive.

Since the Third EAP, environmental protection is seen as having the potential to enhance the competitiveness of the EU's economy. This view displaced the earlier 'limits to growth' argument in favour of a belief in continued economic growth based on environmental protection. As the Commission later stated, 'the main message is that we need to change growth, not limit growth' (CEC 2001a: 16). The Fourth EAP (1987–92) further developed this idea. It drew upon, and helped to promote, the principle of ecological modernization.

> Ecological modernisation is the invention, innovation and diffusion of new technologies and techniques of operating industrial processes, such that economic growth is decoupled from environmental harm.
>
> (Murphy 2000)

Box 6.1 Ecological modernization

Work on ecological modernization grew out of the belief that the decoupling of economic growth from environmental destruction may be an emerging feature of certain advanced industrial economies. Decoupling reduces the amount of physical emissions or natural resource use per unit of economic output. This can be achieved either through increased efficiency, 'eco-efficiency', stemming from technological change, or from a shift to less environmentally harmful products. Decoupling breaks the causal chain between economic activity and negative environmental effects.

Attention is also paid to the benefits of Factor 4 – that is, the argument that the widespread adoption of existing efficient technologies could lead to a quadrupling of energy and resource efficiency. In turn, this would pay for itself through lower consumption of energy. Closely related to this is the idea of the 'triple bottom line'. The triple bottom line is an accounting procedure that encourages corporations to measure their performance not only by the traditional financial bottom line, but also by their social and environmental performance. This is often referred to as 'win–win–win', and has become increasingly fashionable in financial investment management and consulting. It has also found its way into the UK sustainable development strategy, which recommends seeking solutions that provide benefit to all three lines, rather than prioritizing any one over the others.

continued

The founder of ecological modernization theory, Joseph Huber, argued that *super-industrialization* could address environmental problems – that is, the development and application of more sophisticated technologies. Industrial society, he argued, has gone through three phases: industrial breakthrough (1789–1848), construction of industrial society (1848–1980) and a current phase of ecological modernization (1980–). The current phase involves reconciling the impact of human activity with the environment.

The later work of Jänicke and of Simmonis shifted the emphasis from technology to macroeconomic structural change. They argued that ecological modernization involves the restructuring of technologies *and* the sectoral composition of national economies. In other words, ecological modernization is part of a shift, in advanced economies, from energy and resource-intensive industries towards service and knowledge-intensive industries. They argue that this has enabled growth of GDP to be decoupled from energy and resource use.

The third strand of theory focuses less on industry than on the state and its role in the development of ecological modernization policy. The use of the strategy of environmental policy integration is a hallmark of this approach. The Netherlands, Germany and Japan are the leaders in this field. The EU's Fourth EAP is also an example. The approach also seeks to build a new relationship between the state and industry. This has led to the use of a range of new policy instruments, such as voluntary agreements with industry, to manage the environment.

Research by Mol builds upon the sociological theory of risk, to argue that ecological modernization is the reflexive reorganization of industrial society in the face of risk. The restructuring of the Dutch chemical industry is the prime example. The pollution prevention schemes introduced by the US company 3M is another example.

The study of ecological modernization is analytical and descriptive as well as normative. Much of the literature is concerned about what the state and industry *should* do in the pursuit of environmental protection, which will improve economic competitiveness at the micro (within firms) and macro levels.

The ecological modernization approach has provided a new way of thinking, beyond assuming a conflictual relationship between the economy and the environment. It centres on an understanding of the environmental predicament as a problem of efficient resource allocation and use. The key strategy is to achieve eco-efficiency. This perspective regards growth as part of the solution to environmental problems, not as part of the problem.

Sources: adapted from Huber (1982); Simmonis (1989); Jänicke (1992); Sachs (1997); Mol (2000); Murphy (2000).

One association in particular, the World Business Council for Sustainable Development (WBCSD), promotes the principle and practice of ecological modernization. It has helped diffuse these in the business community.

Box 6.2 World Business Council for Sustainable Development

The WBCSD was formed in the run-up to the 1992 Earth Summit, to provide a voice for business at Rio. Its aim is 'to provide business leadership as a catalyst for change toward sustainable development and to promote the role of eco-efficiency, innovation and corporate social responsibility'.

To date, around 170 international companies have joined the WBCSD. It has also a regional network of forty-five national and regional partner organizations, the Business Councils for Sustainable Development.

Members share a commitment to sustainable development through promoting economic growth, ecological balance and social progress. The WBCSD believes that 'the pursuit of sustainable development is good for business and business is good for sustainable development'. The WBCSD specific objectives are:

- to be the leading business advocate on sustainable development;
- to participate in policy development in order to allow business to contribute effectively to sustainable development;
- to demonstrate business progress in environmental and resource management and corporate social responsibility and to share leading-edge practices among its members;
- to contribute to a sustainable future for developing nations and nations in transition.

WBCSD work is structured by several cross-cutting themes: eco-efficiency, innovation and technology, corporate social responsibility, ecosystems, sustainability and markets, and risk. It also produces research reports.

Source: adapted from http://www.wbcsd.org, accessed 6 February 2004.

While ecological modernization theory offers a new, environmentally sensitive way of looking at the relationship between the economy and society, it has none the less been the subject of extensive criticism. The environment is seen to be reduced to concern about resource inputs, waste and pollutant emissions. Critics argue that this reduction has a 'seductive appeal': belief in an efficiency response to the environmental problem minimizes the degree of social and cultural changes that are necessary to promote sustainable development, especially in the high-consumption societies of the West (Weizäcker *et al.* 1997). Nor does this response address the basic ecological contradiction in capitalism – that is, it requires constant expansion of consumption in a world characterized by finite resources.

There has also been criticism of the almost exclusive emphasis on technology and economic entrepreneurs as determinants of social change (Christoff 1996). Social change, especially for Brundtland, is a process involving a broader set of actors and the promotion of sustainable development involves engagement with a deeper set of principles. In particular, the social justice aspects of sustainable development are ignored by this approach (Langhelle 2000).

The empirical focus of the research has also been criticized as overemphasizing the European and Japanese cases. Analysis also ignores the fact that some of the environmental improvements experienced in industrial societies may have come at the cost of displacing environmentally harmful activities to less developed countries (Pepper 1998). Most of Japan's aluminium, for example, is smelted overseas and Japanese forests are intact, as practically all timber is imported (Goodland and Daly 1996).

Nevertheless, ecological modernization theory has been important for several reasons. First, it provides a way of dealing with the evidence that suggests that advanced industrial countries have shifted from an earlier phase of crude, environmentally damaging industrial capitalism to a phase of making progress in dealing with some environmental problems (Murphy 2000), as is represented by the Kuznets curve. Second, it helps in theorizing the changes in style and content of European environmental policy, from an earlier 'command and control' approach to one that makes more use of the precautionary principle, environmental policy integration, and integrated pollution prevention, and of voluntaryism and market-based incentives. Ecological modernization would appear to straddle the weak and strong versions of sustainable development. Third, ecological modernization has also been important as an economic strategy. It has helped the promotion of eco-efficiency and environmental benefits. Acceptance of the principle has encouraged industry to be more resource-efficient and has allowed environmental protection goals to be positively linked with economic development strategies. This was particularly important at the EU level. As this chapter progresses it will become clearer how the new way of thinking provided by the ecological modernization argument helped to shape the characteristics of EU sustainable development policy. While ecological modernization has helped industry to contribute to environmental well-being through changes in the production process, consumption also needs to be addressed. The last few decades have witnessed the development of groups promoting ecologically responsible consumption – for example, through the purchase of Green consumer products and organic food, and through recycling. However, Green consumerism has yet to address both the patterns *and* levels of consumption and to promote ecologically responsible citizenship among consumers.

Setting sustainable development objectives

The Fifth EAP, *Towards Sustainability* (1993–2000), made the first explicit policy commitment to the promotion of sustainable development in the EU (CEC 1992). Acknowledging the influence of Brundtland, the Fifth EAP defines sustainable development as continued economic and social development without detriment to the environment and natural resources, on the quality of which continued human activity and further development depend (CEC 1992). It also called for the use of a wide range of policy instruments, including fiscal and voluntary measures, as promotion tools.

Two immediate pressures influenced the EU's commitment. The first was recognition of the increased burdens that European integration, particularly the completion of the Single European Market, posed for natural resources, the environment and the quality of life within the Union. The Single European Market, for example, increases trade between member states, putting more pressure on the environment from the transport sector. This required an approach that would reconcile the tensions between deepening European economic integration and the EU's ever growing environmental agenda.

The second pressure stemmed from international engagement, particularly the Community's participation in the UNCED process. The Fifth EAP was drawn up in parallel with preparations for the Rio Earth Summit. Action taken under the Fifth EAP came to represent the EU's main strategic response to the obligations it incurred at Rio in the period up to 2001. When viewed from this international perspective, it is worth noting that the EU commitment to sustainable development was also driven by a sense of moral obligation, in particular acceptance of the principle of common but differentiated responsibilities. As the Commission has argued:

> As Europeans and as part of some of the wealthiest societies in the world, we are very conscious of our role and responsibilities. [A]long with other developed countries, we are major contributors to global environmental problems such as greenhouse gas emissions and we consume a major, and some would argue an unfair, share of the planet's renewable and non-renewable resources.
>
> (CEC 2001b: 11)

The Fifth EAP is a practical, policy-orientated document which sets five objectives for the promotion of sustainable development (Box 6.3).

Box 6.3 Key EU objectives for the promotion of sustainable development

- *Development of strategies in seven environmental priority areas*: climate change, acidification, biodiversity, water, urban environment, coastal zones and waste.
- *Targeting five key sectors* of environmental policy integration: industry, energy, transport, agriculture and tourism.
- *Broadening the range of instruments* used to promote sustainable development so as not to rely exclusively on legislation: this includes the use of fiscal, market and voluntary policy tools, and the involvement of actors other than regulatory authorities.
- *The application of the principle of shared responsibility*: the promotion of sustainable development is the responsibility of all the different levels of governance (EU, member state, regional and sub-national). It also requires new forms of partnership between economic and social actors and public policy makers.
- Deepening the Community's *international engagement*, particularly with respect to global environmental issues.

Source: adapted from CEC (1992).

The Fifth EAP has been subject to extensive reviews, including by the Commission and the European Environment Agency. However, evaluation of the Fifth EAP has not been an easy task. At the general level, the lack of clarity in the term 'sustainable development', which the Fifth EAP is designed to promote, made it difficult to measure progress. This led to the construction of sustainable development indicators by the European Environment Agency to address the problem. Furthermore, the Fifth EAP did not always set clear implementation targets – for example, it set no quantitative targets for the manufacturing sector. To add to these difficulties, the EAPs are non-binding policy documents and, as such, member states have no legal obligation to report on their implementation.

Nevertheless, it has been possible to reach a consensus on EU efforts to promote sustainable development through the Fifth EAP: while there has been some decrease in the negative pressures exerted on the environment, progress has been very limited.

> The European Union is making progress in reducing certain pressures on the environment, though this is not enough to improve the general quality of the environment and even less to progress towards sustainability. Without accelerated policies, pressures on the environment will continue to exceed human health standards, and the often-limited carrying capacity of the

environment. Actions taken to date will not lead to full integration of environmental considerations into economic sectors or to sustainable development.

(EEA 1995: 1)

There has been no reversal in economic and social trends that are harmful to the environment, particularly in relation to transport, energy and tourism. Clearly, environmental policy integration remains weak. More substantively, production and consumption trends have remained unchanged since the introduction of the Fifth EAP (EEA 1995). In global terms, the record also shows depressing trends. The EU is responsible for 15–20 per cent of the world's resource consumption: this balance has remained unchanged for the last thirty years, almost the entire period during which the EU has been involved in environmental management. Faced with these problems, the Commission's *Global Assessment* concluded that 'unless more fundamental changes are made, the prospects of promoting sustainable development remain poor' (CEC 2000b).

The Sixth EAP, *Our Future, Our Choice* (2001–10), attempts to address some of these shortcomings (CEC 2001b). The Sixth EAP focuses on four priority areas (Box 6.4). The Sixth EAP makes environmental policy integration one of five key 'thematic strategies', alongside more effective policy implementation, enhanced citizen and business engagement, and developing a more environmentally conscious attitude to land use. There are some new connections made in the Sixth EAP, especially between health and the environment. The Sixth EAP calls for the full application of the precautionary principle, especially when it comes to the impact of poor environmental quality on the health of vulnerable groups, such as children and the elderly. The Sixth EAP also calls for the decoupling of resource use from growth.

Box 6.4 Priorities of the Sixth EAP

- Climate change.
- Protecting nature and biodiversity.
- Environment and health.
- Resource and waste management.

In addition to framing the promotion of sustainable development within EAPs, two strategy documents structure EU sustainable development policies. The first is the *Biodiversity Strategy* (CEC 1998b), which represents the Community's response to the obligations incurred under the CBD. The second strategy document, *A Sustainable Europe for a Better World: A European Union Strategy for Sustainable Development*, was introduced in preparation for the WSSD,

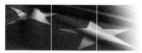

Environment 2010:
Our Future, Our Choice

6th EU **Environment** Action Programme

2001–2010

European Commission

Figure 6.1 *Promoting sustainable development in high-consumption societies requires environmentally responsible choices, about consumer lifestyles, production goals and institutional responsibility*

Courtesy: Office for Official Publications of the European Communities

Johannesburg, 2002 (CEC 2001c). This aims to 'establish a long-term strategy to dovetail policies for economically, socially and environmentally sustainable development'.

The Sustainable Development Strategy builds on three core themes. The first, called 'cross-cutting proposals', is aimed at greater policy consistency and ensuring that policies give priority to sustainable development. It has led, for example, to the requirement that any new legislative proposals must include an assessment of their potential economic, environmental and social costs and benefits. The second theme, known as 'measures to attain long-term objectives', includes strategies to limit climate change. The climate change strategy aims to ensure that the EU meets the commitments it has made under the Kyoto Protocol, and it hopes to follow this by reducing greenhouse gas emissions by an average of 1 per cent per year over 1990 levels up to 2020. Limiting major threats to public health is another long-term objective, especially in relation to food safety, chemicals and resistance to antibiotics. More responsible management of natural resources is also an objective. Here the Commission has the ambitious target of breaking the link between economic growth and the use of resources and halting the loss of biodiversity by 2010. Reducing regional disparities is another long-term objective, as is the development of environmentally friendly transport systems. With such ambitious targets, it is not surprising to discover that DG Environment, the administrative branch of the Commission responsible for environmental matters, has adopted a 'gradual approach', based especially on meeting the Kyoto Protocol obligations.

The third theme, progress reviews, involves a review at each spring European Council meeting to check on progress in implementing the Sustainable Development Strategy. This lends high-level political weight to the review process. Progress will also be subject to review at the beginning of each Commission's term of office. The Commission also plans to give stakeholders a chance to have their say, by organizing a Stakeholder Forum every two years to assess the strategy.

Together, these EAP and strategy documents represent the policy framework of the EU, within which it puts into practice its legal obligation (declaratory intent) to promote sustainable development. They frame the context within which action, secondary legislation, specific programmes and funding are structured. A special fund has been set aside to finance this commitment, known as the Lending Instrument for the Environment (LIFE) Fund. LIFE provides financial support for environmental and nature conservation projects throughout the EU, in candidate countries and in bordering regions.

Distinct EU understanding of sustainable development

Analysis of its EAPs and environmental strategy documents can reveal a great deal about the EU's understanding of sustainable development. At a general level, such analysis reveals that, throughout the First EAP, the imperative of economic growth continued to take precedence over environmental protection. By the Fifth and Sixth EAPs, however, the Union had evolved an environmental policy that intertwined the twin imperatives of economic development and environmental protection in a new way: they have become compatible, mutually reinforcing aims of policy. Through this historical progression, a distinct EU understanding of sustainable development has emerged (Box 6.5).

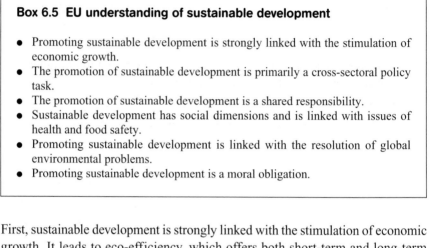

Box 6.5 EU understanding of sustainable development

- Promoting sustainable development is strongly linked with the stimulation of economic growth.
- The promotion of sustainable development is primarily a cross-sectoral policy task.
- The promotion of sustainable development is a shared responsibility.
- Sustainable development has social dimensions and is linked with issues of health and food safety.
- Promoting sustainable development is linked with the resolution of global environmental problems.
- Promoting sustainable development is a moral obligation.

First, sustainable development is strongly linked with the stimulation of economic growth. It leads to eco-efficiency, which offers both short-term and long-term competitive advantages to European industry. This provides a way of decoupling economic growth from environmental degradation. This understanding is closely aligned with the ecological modernization approach.

Second, the promotion of sustainable development is primarily a cross-sectoral policy task. It acknowledges that environmental pressures, stemming, for example, from transport, energy or agriculture, often outweigh the benefits of new regulations. This has led to a management approach that combines regulation with other forms of steering, such as the use of voluntary agreements.

Third, the promotion of sustainable development is a shared responsibility, a characteristic arising partly because the EU is a system of multi-level governance. This means that all levels, from the EU to the sub-national, regional and local levels, have to play a role if policy is to be successful. It also enables the Union to align its practices with the principles of good environmental governance (such as subsidiarity and participation) developed since Rio. However, it is also a shared responsibility, because promoting sustainable development is an on-going process of stimulating behavioural and normative changes. This involves social and economic actors to acknowledge the environmental consequences of their behaviour and to modify their actions accordingly. Here the concept of environmental citizenship becomes important, where citizenship not only brings environmental rights (clean air, health and safety) but environmental obligations. Unfortunately, action in this area remains limited, especially when judged by the efforts to adopt sustainable consumption patterns within the high-consumption societies of Western Europe.

Fourth, sustainable development has social dimensions, where strong commitment to social concerns has always marked European political culture, albeit manifested in different ways in different member states. In the Sixth EAP this is linked with issues of health and safety, especially food safety.

Fifth, sustainable development is linked with the resolution of global environmental problems, in particular climate change and biodiversity.

Finally, sustainable development is a moral obligation, not only because the Commission has recognized that European economic development has made a major contribution to the current global environmental crisis, but also because Europe is seen as having the capacity to lead the way in resolving that crisis.

The characteristics of the EU's understanding of sustainable development bear resemblances to the understanding formulated by Brundtland. These include a concern for the global dimension, acknowledgement of the differentiated responsibilities of the North and the recognition of the social dimensions of sustainable development. However, the European approach prioritizes European concerns, neglecting to embed these in the wider global context. First, the traditional emphasis on economic growth is maintained. Less attention is given to issues of development, in particular what the Brundtland call for necessary social and economic improvement in the Third World might mean for European growth aspirations. Second, the understanding of what constitutes *global* problems is limited to areas that are having a direct impact in Europe, in particular climate change and biodiversity loss. Third, discussion of the social aspects of sustainable development increasingly stresses issues of concern to Western consumers, such as food safety and health, to the exclusion of issues such as food security. Absent from this discussion are issues of inter-generational and intra-generational equity, particularly in relation to the need to modify the high consumption and resource-use patterns of Western consumers. The idea that the promotion of sustainable development can rely upon partnership is premised upon the existence of a developed civil society and of responsible entrepreneurship, or at least an economic sector that has come to accept the benefits of ecological modernization. This idea has not taken hold at the global level and, as discussed above, may not be capable of having a global remit.

The task ahead: achieving environmental policy integration within the European Union

One of the key challenges facing the EU is to integrate environmental considerations into its other policies, particularly at the sectoral levels. The achievement of environmental policy integration presents difficulties at several levels for the EU. At the institutional level, it impinges upon policy sectors managed by different Directorates-General (DGs) within the Commission. Unfortunately, the DGs find it difficult to see beyond the limits of their own competences, which they often guard jealously. Consequently, this pits DG Environment against many of the other DGs (Wilkinson 2002). In addition, within the DGs there is a widespread belief that the promotion of sustainable development is the business of those who deal with the environment.

Box 6.6 Environmental policy integration

At the simplest level, environmental policy integration means the integration of environmental considerations in the design and implementation of policy. However, this provides only a very general guide for policy makers. It merely points to the need for policy coordination and the removal of contradictions between policies. This, however, can reduce the challenge to little more than a search for good policy practices: all good policy making should attempt to achieve a high level of policy coordination. It does not provide guidelines for dealing with the real conflicts of interest that arise with respect to environmental and economic issues. As the principle is poorly defined, it has taken on different meanings when used by different policy actors. In the EU, for example, environmental policy integration often means the achievement of 'balance', not reconciliation, between economic and social priorities.

Lafferty and Hovden provide a comprehensive understanding of environmental policy integration:

- the incorporation of environmental objectives into all stages of policy making in non-environmental policy sectors, with a specific recognition of this goal as a guiding principle for the planning and execution of policy;
- accompanied by an attempt to aggregate presumed environmental consequences into an overall evaluation of policy, and a commitment to minimize contradictions between environmental and sectoral policies by giving principled priority to the former over the latter.

Lafferty and Hovden argue that *principled prioritization* of environmental policy objectives forms the core of environmental policy integration. To argue that sustainable development merely implies that the essential needs of the world's poor and of future generations (the two key target groups of Brundtland) should be

---

'balanced' with a myriad of other societal goals misses, in their view, the fundamental premise of the Brundtland Report and its follow-up within the UNCED process.

The approach of Lafferty and Hovden makes the principle of environmental policy integration a fundamental challenge to traditional models of development. Requiring a principled and consequential integration of environmental considerations into all sectoral activity, it involves a significant break with the traditional capitalist economic development model. This, in turn, points to the radical nature of the sustainable development agenda.

Source: Lafferty and Hovden (2003).

At the instigation of the Swedish Prime Minister in 1997, new efforts were made to integrate environmental concerns into EU policies, launching the so-called Cardiff Process. The process is overseen by the European Council – that is, the heads of state and government of the member states – and works through the Council of Ministers – that is, the nine groupings of ministers from the member states, such as the Agriculture Council and the Transport Council.

Box 6.7 The Cardiff Process

The Cardiff Process is aimed at achieving environmental policy integration at the sectoral level. Launched in 1998, it called upon the nine councils to develop environmental policy integration strategies. The councils were also asked to develop mechanisms, based on indicators, for reporting their progress. The European Environmental Agency helps with the development of indicators for measuring progress.

All but one of the nine council groups has submitted integration strategies. The Fisheries Council has not submitted a strategy, because of on-going difficulties with the Commission about reforming the EU Common Fisheries Policy.

The Cardiff Process has set in progress a cycle for reviewing progress in sectoral integration at each of the European Council meetings. Despite the flurry of communications and policy documents that have accompanied the Cardiff Process, its achievements have been limited. There has been a great deal of unevenness in the response of the different councils to the Cardiff Process, with DG Agriculture having made the most progress and DG Internal Market and DG Fisheries among the least.

Source: adapted from Kraemer (n.d.).

The EU uses a variety of policy tools to promote environmental policy integration (Box 6.8). The Union faces difficulties in trying to realize its commitment to environmental policy integration. This can be seen more clearly by reviewing the key policy sectors that have been singled out for integration measures, both in the EAPs and in the Cardiff Process.

Box 6.8 Tools for environmental policy integration

- *Coercive measures:* banning or restricting certain activities or the sale of certain goods that are harmful to the environment. For example, the Commission is proposing a ban on phthalates in children's toys.
- *Regulatory measures:* passing legislation, setting standards, targets and licensing to control the activities of particular sectors. For example, the EU directive (2000/76/EC) on the incineration of waste requires incineration plants to publish annual reports, including information on emissions.
- *Fiscal measures*, including taxation, tax breaks or subsidy reform, which change the price signals in the market place in favour of more environmentally friendly production. For example, the Irish government introduced a tax on plastic bags from supermarkets.
- *Voluntary measures*, including voluntary agreements with economic sectors and the adoption of environmental management systems.
- *Information measures:* campaigns to raise awareness and help consumers make informed choices, for example promoting energy-saving measures, household waste recycling and the purchase of environmentally friendly products.
- *Assessment measures*, including specific tools such as environmental assessment, for example the introduction of Strategic Environmental Assessment.

Source: adapted from European Environmental Agency (2002a: 17).

Agriculture

The Common Agricultural Policy (CAP), in modernizing European agriculture, has also led to serious environmental deterioration. The high level of support it gives to maintain agricultural prices has encouraged intensive forms of agriculture. The resultant increased use of fertilizers and pesticides has polluted water and led to soil contamination. It has also led to the destruction of some important ecosystems through the removal of hedges, stone walls and ditches, and the drainage of wetlands. This has reduced natural habitats for a large number of birds, plants and other forms of wildlife. In some regions, intensification has resulted in over-consumption of water and has speeded up soil erosion.

The 1992 CAP reform attempted to deal with these environmental problems. The reforms encouraged less intensive production and the reduction of surpluses and introduced agri-environmental and forestation programmes. The second CAP reforms of 1999 led to a further strengthening of agri-environmental measures. The Agenda 2000 reforms, introduced in preparation for the eastern enlargement of the EU, also sought to strengthen environmental measures (CEC 1997c, 1999a). These allow payments to farmers who voluntarily undertake environmental work over and above good agricultural practice. In addition, member states may link direct payments to farmers with their compliance with environmental requirements. These new rules are designed to reduce payments to farmers that do not comply with EU environmental legislation (CEC 1999a). However, while there has been some reduction in the use of fertilizers and pesticides, nitrate pollution and eutrophication remain serious (EEA 2002a). Emissions of ammonia, a greenhouse gas, from the agricultural sector also remain a problem.

Despite the environmental reforms of the CAP, the promotion of sustainable agriculture and, more generally speaking, of sustainable forms of rural development remains hampered by several major problems. Member states retain discretion over how to implement environmental measures in the agricultural sector and the level of commitment varies from one member state to another. Poor compliance with environmental legislation also plagues the agricultural sector. While the EU hopes that its Biodiversity Strategy in particular will promote sustainable agriculture, there are continuing problems with the implementation of key pieces of law on biodiversity, including the Birds and Habitats Directives and the Natura 2000 programme (Baker 2003). More generally, however, the CAP is still driven by an approach that prioritizes output-orientated, intensive production in the agricultural sector, including in the new member states of Eastern Europe.

Energy

The energy sector has made some progress in integrating environmental considerations into policy (CEC 1998c). The need to meet the UNFCCC obligations has resulted in institutional capacity building and the development of an EU Energy Framework Programme (1998–2002), as well as a new emphasis on the promotion of renewable energy and energy efficiency. Environmental legislation on large combustion plants and policy developments in relation to combined heat and power production (co-generation) and on the disposal of disused offshore oil and gas installations represent other developments.

While this may appear a laudable set of initiatives, developing an energy policy response to climate change remains one of the most pressing problems facing the Union and provides one of the clearest examples of the importance of environmental policy integration. Since 1994, carbon dioxide (CO_2) emissions have been increasing in the EU and in most individual member states. Emission reductions up to 2002 were insufficient to enable the Union to meet its Kyoto targets. Meeting those targets has now become dependent upon emissions trading and the introduction of measures to help developing countries improve their emission levels (CEC 2004b). These strategies do little to force a direct reduction in emission levels in member states. In addition, several member states will in fact exceed their legally binding targets during this period. There are also marked differences between the contributions of different economic sectors. In the transport sector CO_2 emissions are expected to rise by 39 per cent by 2010, compared with 1990 levels. In the energy sector, emissions should stabilize, but a 4 per cent increase in emissions from households and the tertiary sector is expected in the next few years. In contrast, CO_2 emissions from the industrial sector should fall by 15 per cent between 1990 and 2010.

Box 6.9 The Kyoto sectoral challenges

The states signatory to the Kyoto Protocol have undertaken to reduce emissions of six greenhouse gases by 8 per cent between 2008 and 2012, with an interim target set for 2005. To meet these targets, initiatives in the areas of transport, energy, agriculture and industry are required:

- within the energy sector, the use of renewable energy needs to increase, as does energy efficiency;
- in relation to transport, there is an urgent need to reduce emissions from passenger cars, to improve transport pricing and to enhance the development of rail transport. These, in turn, require changes in land-use policy, planning and urban design as well as lifestyle patterns;
- the agricultural sector still needs to enhance reforestation, develop better practices in relation to land management, biodiversity preservation and livestock feeding regimes, and reduce the use of fertilizers in crop production;
- industry still needs to promote innovation in the field of clean technology and increase eco-efficiency.

Sources: adapted from CEC (1999b, 2001d).

Industry

In the industrial sector, many large European companies have already reaped the rewards of eco-efficiency, through their ecological modernization strategies. This includes the introduction of an integrated product policy and of environmental management and audit systems. Industry also represents one of the key sectors where the Commission has been able increasingly to make use of the principle of shared responsibility, especially in the development of voluntary agreements. The Auto Oil Programme, a technical programme designed to reduce emissions, in particular from road transport, so as to improve air quality, provides good examples of the use of voluntary agreements.

However, progress is limited, especially in small and medium-sized, as well as domestically oriented, businesses. There are also serious weaknesses in the use of the strategy of environmental policy integration in this sector as a tool for the promotion of sustainable development. The gap between the ecological modernization of the industrial sector and the promotion of sustainable development has already been discussed. In addition, the introduction of eco-efficiency measures does little to promote sustainable consumption patterns. There is a need to develop environmentally responsible entrepreneurship, to increase the awareness of industry and to encourage change in the behaviour of consumers. Such changes are hard to promote, as they impinge upon consumption choices and question the need for much of what is produced, albeit with greater eco-efficiency, by European industry.

Fisheries

The EU developed its Common Fisheries Policy (CFP) because of disputes between member states over territorial fishing rights and to prevent overfishing. It has introduced annual quotas on the take of Atlantic and North Sea fish and regulations on fishing areas and equipment, including limits on the mesh size of fishing nets and on the size of fish caught.

In 2000 the Commission set about finding new ways of integrating wider nature conservation objectives into the CFP. This led to the 2002 Community Action Plan, aimed at integrating environmental protection requirements into the CFP (CEC 2002a). Furthermore, the Biodiversity Strategy also deals with the development of a sustainable fisheries policy, through its specific Biodiversity Action Plan for Fisheries.

However, overfishing remains a major problem. Fishing methods still need to be improved to reduce discards, incidental by-catch and impact on habitats. The aquaculture sector (fish farming) is also a source of environmental pressure and the EU still needs a strategy for distant-water fisheries. The member states also need to take action. They still have to meet the obligations imposed on them by EU nature protection legislation (the Birds and Habitats Directives) and to eliminate state aid likely to increase fishing capacity. The outlook for efforts to integrate environmental considerations into the Common Fisheries Policy remains pessimistic. A strategy for the conservation and sustainable use of commercial stocks and marine ecosystems has yet to be developed, and EU-level efforts to date have not halted the decline in fish stocks.

Transport

EU transport policy has long been a key source of environmental stress, particularly the large-scale infrastructure projects that were introduced to help the completion of the European single market. Because of the growth in transport, and the shift to road and aviation, CO_2 emissions from the transport sector are continuing to grow. Transport is the fastest-growing energy consumer in the EU. Unfortunately, while there have been some, albeit slight, improvements in the energy efficiency of passenger transport, there has been no similar improvement in freight transport (EEA 2002b).

A lack of commitment in many member states also remains a problem. This is especially true in peripheral areas and in new member states from East and Central Europe, where road building represents a strategic response to the competitive challenges posed by the completion of the European single market. This is discussed in more detail in Chapter 8.

Conclusion

It would appear that the EU is making progress in reducing certain pressures on the environment. A comprehensive range of legislation, strategy documents and action programmes now frames EU environmental management. However, this is not enough to improve the general quality of the environment and even less to promote sustainable development. Environmental policy integration remains a daunting challenge. Without accelerated efforts and commitment, including on behalf of both producers and consumers, pressures on the environment will continue to exceed the limited carrying capacity of the environment.

Thus it would appear that when analysis moves from exploration of the Union's constitutional and declaratory commitments to its implementation efforts, especially at the sectoral level, a different, and altogether more pessimistic, picture emerges of the promotion of sustainable development in the EU. The eastern enlargement of the Union is likely to see this capability–expectation gap (Hill 1993), the gap between policy outcome and declaratory intent, grow ever wider.

These negative reviews, however, do not mean that the EU commitment to the promotion of sustainable development has been a policy failure. Rather, its commitment is important for several reasons.

First, the EU has set out an ambitious vision of sustainable development for Europe. Its acceptance by the EU is of deep symbolic importance. In addition, legal, treaty-level obligations back this vision. The Commission has an on-going duty to ensure that these obligations are met. As the official 'guardian of the treaty', the Commission takes this duty very seriously.

Second, the term 'sustainable development' now permeates the official discourse of the Union. In its official programmes, legal commitments and public discourse, the EU has moved from an earlier phase characterized by tension between its different policies to a new phase of learning how to achieve policy resolution.

Third, medium-term policy frameworks and strategies (the Fifth and Sixth EAPs, the Sustainable Development Strategy, the Biodiversity Strategy) have been put in place, and a high-level policy process (the Cardiff Process) has been launched. Policy initiatives have also been forthcoming (sector-specific integration strategies, biodiversity Action Plans). This shows clear evidence of policy learning, as seen by the willingness of DG Environment in particular to expose its policies to on-going evaluation, and its search for new and improved ways of putting its commitment into practice.

Fourth, even if policies fall far short of promoting sustainable development (and they most certainly will), the EU commitment provides an important environmental and development criterion against which the integration process can in future be appraised. A failure to realize that vision in the period of the last two EAPs (1993–2003) was to be expected, not least because the promotion of sustainable development is a long-term process of social, cultural, political and economic change. What is important is that the EU has launched Europe on a *path* towards sustainable development.

Finally, it is also important in shaping the negotiating position and behaviour of the EU at the international level: sustainable development has become a norm of EU policy, especially at the international level. Beyond the borders of the Union, the commitment can also help to shape the EU's identity by marking it out as *different* from other actors. It marks a major difference from the US, which is reluctant to address key sustainable development issues, in particular climate change.

Summary points

- Declaratory commitment to the promotion of sustainable development is high in the EU. This is important, as the power of ideas in politics should never be underestimated.

- For the EU, the promotion of sustainable development provides a way of decoupling economic growth from environmental destruction. It is seen primarily as a technical, managerial task of decoupling through eco-efficiency. This understanding of sustainable development does not challenge the Western economic development model, as it aims neither to limit growth nor to change existing patterns of high consumption.
- Ecological modernization arguments have played a key role in ensuring the acceptance of environmental norms within the EU. While there are different strands in ecological modernization theory, they all conceive environmental deterioration as a challenge requiring and forcing socio-technological and economic reform. In addition, there is emphasis on developing modern institutions as carriers of ecological restructuring, such as the market, the scientific and technological communities and the state. However, while the literature often confuses ecological modernization with sustainable development, ecological modernization is a more limiting concept that does not address the underlying contradictions in capitalism: a logic of ever-increasing consumption in a world characterized by material resource limitations.
- International developments within UNCED strongly influenced the EU commitment to sustainable development. However, the EU's understanding of sustainable development differs in fundamental respects from that presented by the Brundtland Report.
- Efforts to date by the EU are not sufficient to promote sustainable development. The EU faces considerable challenges in implementing its legal and declaratory commitment to sustainable development. It is committed to the use of the strategy of environmental policy integration. Use of this strategy to date, while stimulating good governance practices, is not addressing the fundamental challenges posed by the commitment to promote sustainable development.

Further reading

Critical commentary on EU sustainable development policy

Baker, S. (2000) 'The European Union: integration, competition, growth – and sustainability', in W.M. Lafferty and J. Meadowcroft (eds) *Implementing Sustainable Development: Strategies and Initiatives in High Consumption Societies*, Oxford: Oxford University Press, 303–36.

Jordan, A. (ed.) (2002) *Environmental Policy in the European Union: Actors, Institutions and Processes*, London: Earthscan.

Ecological modernization

Hajer, M. (1995) *The Politics of Environmental Discourse: Ecological Modernization and the Policy Process*, Oxford: Oxford University Press.

Langhelle, O. (2000) 'Why ecological modernization and sustainable development should not be conflated', *Journal of Environmental Policy and Planning*, 2: 303–22.

Murphy, J. (2000) 'Ecological modernisation', *Geoforum*, 31: 1–8.

Sachs, W. (1997) 'Sustainable development', in M. Redclift and G. Woodgate (eds) *The International Handbook of Environmental Sociology*, Cheltenham: Edward Elgar, 71–82.

Environmental policy integration

Lafferty, W.M. and Hovden, E. (2003) 'Environmental policy integration: towards an analytical framework', *Environmental Politics*, 12(3): 1–22.

Web resources

http://europa.eu.int (official EU portal).
http://wbcsd.ch (Web site of WBCSD).

7 Challenges in the Third World

Key issues

- **The development agenda; reconceptualizing development.**
- **Least developed countries.**
- **Women, environment and development.**
- **Science and knowledge.**
- **Free trade and the WTO; Global Environment Facility; World Bank.**
- **Financing sustainable development: Monterrey Consensus.**

One of the most pervasive features of UNCED discussions, documents, meetings and Summits is the attention given to the environmental dimensions of the division between the North and the South, between the rich and the poor, between the high-consumption societies of the industrialized world and those in the Third World struggling to sustain their basic livelihoods. The Brundtland Report and the related UNCED process acknowledged that both the positive and the negative consequences of industrialization and agricultural modernization, when viewed from a global perspective, are inequitably distributed. In addition, the consequences of this type of development for global environmental change, including climate change and biodiversity loss, are uneven in their impact.

This chapter looks at the problems facing Third World countries, and how these are shaping their emerging environment and development agendas. It begins by exploring what 'development' has meant for the Third World and outlines new efforts to reconceptualize development from a Third World perspective. Attention is then turned to the relationship between the industrialized countries and the Third World within UNCED. This exposes the contrast between the approach taken by UNCED and the analysis presented by Brundtland. Five key themes in relation to the promotion of sustainable development in the Third World are reviewed in detail. This draws attention to the structural barriers that exist at the international level, including international funding regimes.

Critique of the development agenda

Many theorists argue that the Western development model sustains inequalities and leads to underdevelopment in the Third World. In the decades since the Second World War, many Third World countries have paid a high economic, social, environmental and cultural price for adopting policies aimed at 'catching up' with Western development. Among rural populations in particular, the Western model is seen as having undermined traditional subsistence agriculture and directed resources towards the production of cash crops and away from traditional food crops. While the former command ever-decreasing prices on the world market, the shortage of the latter results in on-going crises of food insecurity and hunger (Shiva 1989).

Many critiques of development have drawn upon theoretical perspectives of imperialism and colonialism. Imperialism has been conceptualized in different ways, but describes 'theories and practices developed by a dominant metropolitan centre to rule distant territories, by force, by political means or by economic, social, and cultural dependence' (Banerjee 2003: 146). Colonialism, a consequence of imperialism, involves the establishment of settlements in outlying territories. A range of relations, between nation states, international institutions and transnational corporations, structure the process. From this perspective, global institutions, such as the Global Environment Facility, the World Bank and the World Trade Organization (WTO), form an important way of institutionalizing the relationships of imperialism and colonialism. This allows, for example, colonial relations to be played out in trade conflict between the industrialized countries and the Third World. As such, the development agenda becomes part of the way in which the 'South' is constructed as being in need of 'development' and 'progress', preferably achieved through the transfer of Western science and technology.

> Ever since President Harry S. Truman coined the notion of 'underdevelopment' in his inaugural address in January 1949, and promised assistance to the countries of the Southern hemisphere in their efforts to catch up with the North, it has been taken for granted that, first, development could be universalized in space and, second, that it would be durable in time. This belief has proved wrong. Development has in fact . . . deepened the crisis of injustice between North and South, just as it has provoked a manifold crisis of nature which undercuts its prospects for the future. It has revealed itself to be finite in (global) space as well as in time.
>
> (Sachs 1997: 71)

This perspective on development has been used to criticize the promotion of sustainable development, as embedded in the UNCED process. The discourse on sustainable development is seen to share characteristics of colonizing discourse, becoming another example of 'a Western style for dominating, restructuring, and having authority over' the Third World (Said 1979: 3). Rather than reshaping markets and production processes to fit the logic of nature, critics argue that sustainable development is a management discourse that allows the logic of markets and capitalist accumulation to determine the future of nature (Shiva 1991). 'In the sustainable development discourse, poverty is the agent of environmental destruction, thus legitimating prior notions of growth and development' (Banerjee 2003: 159). Such critics, rather than promote sustainable development, call for a radical reconceptualization of development itself (Box 7.1).

Box 7.1 Reconceptualization of development

- Acknowledging the structural power relations that lie at the heart of our environmental crisis.
- Recognizing the structural and natural limits of sustainable development.
- Moving beyond managerial efficiency to also incorporate new critiques of modernity, and its 'meta-narratives' of progress.
- Reversing the industrial appropriation of nature.
- Searching for alternatives to development to restructure the system of political economy.
- Shifting the focus from capital and markets to achieve sustainable development by developing new ways of thinking and knowing.
- Applying insights from the full variety of knowledge and using them to challenge existing views of the world and nature.

Sources: adapted from Lefebvre (1991); Escobar (1995); Redclift (1997); Banjeree (2003).

However, adopting a Third World perspective does not necessarily lead to the abandonment of the sustainable development project, especially as formulated by Brundtland. Development remains a prime objective of Brundtland, where the starting point of analysis is that, in the face of poverty, disease and hunger, it is neither ethically nor politically acceptable to expect the Third World to halt its development. Indeed, argued Brundtland, such is the overwhelming need of the world's poor that economic development may take precedence over environmental protection considerations. This argument also recognizes that economic growth can provide the resources necessary to protect the environment in the poorest countries, not least because *lack* of resources causes as much environmental damage as do the pressures of industrialization. Nevertheless, as discussed in Chapter 2, while the international, institutional engagement with this concept can be criticized, at the heart of the Brundtland formulation there remains a strong argument for the radical transformation of the structures of political and economic power. In this transformation lies the scope for Third World communities to construct varied and relevant development paradigms that reflect their needs, values and aspirations.

> A baby born in the United States represents twice the impact on the Earth as
>   one born in Sweden
> 3 times one born in Italy
> 13 times one born in Brazil
> 35 times one in India
> 140 times one born in Bangladesh or Kenya
> 280 times one born in Chad, Rwanda, Haiti or Nepal.
>
> (Ehrlich and Ehrlich 1989)

In addition, Brundtland also recognized that, given the earth's limited resources, bringing the low-income countries up to the affluence levels found in OECD countries is a very unrealistic goal. One way in which the impact of human activity on the environment can be reduced is by changing production. This would involve having more high-value, low-throughput production, a feature of ecological modernization, discussed in Chapter 6. Nevertheless, Brundtland points out that global equity, at current OECD consumption levels, is simply not possible: 'Present patterns of OECD resource consumption and pollution cannot possibly be generalised to all currently living people, much less to future generations, without liquidating the natural capital on which future economic activity depends' (Goodland and Daly 1996: 1004).

The promotion of global sustainable development involves a twofold task: overcoming the barriers to sustainable development in the developing world and reducing the high consumption levels in the industrialized world. This is what is meant by the claim that sustainable development requires the principle of equity to be inserted into the development paradigm. An equitable development paradigm addresses the inequalities wherein the industrialized countries overconsume while most of the rest of the world consumes at barely subsistence level.

The Third World within UNCED

It is fair to say that most of the Third World approach UNCED with a mixture of fear and hope (Grubb *et al.* 1993). The fear is that the people of the South are to sacrifice their chance to live 'the American dream', while having to withstand the worst of the environmental and social costs of the economic growth of the industrialized world. There is also fear that the North is using its environmental concerns to place new conditions upon overseas development aid (ODA). Such 'Green conditionality' may well reflect Northern priorities and perceptions and be supervised by global financial institutions in which the South has little faith. Behind this fear lies an understandable feeling of resentment: the international environmental agenda is another example of Western arrogance towards the Third World.

Yet there is also hope. Here lies the possibility that the environmental concerns of the North may give the South real political advantage in global politics. If the North wants the South to change its future behaviour and development paths, then it will have to meet Southern demands (Grubb *et al.* 1993). These demands include debt relief, increased ODA as well as increased market access and prices for their commodities.

For its part, the North feels that the South has to make some changes. Chief among these is the need to curb population growth and to introduce more stable political structures to manage better its civil unrest. Most Third World countries resist international efforts to discuss population targets and controls on population growth. The UN has found it difficult to find a common ground on which to discuss this topic. It raises sensitive cultural and religious issues and touches upon personal matters of reproduction. Yet population stability is an essential element in the promotion of sustainable development, not least because unregulated population growth may push us beyond the carrying capacity of the planet, while simultaneously undermining efforts to raise living standards for the world's poor. However, while population stability is essential to reduce environmental pressures, a distinction needs to be made between population growth and population impact: in the South, population growth is the higher; but the population of the North has the greater environmental impact. Overconsumption in OECD countries contributes more to global unsustainable patterns of development than population growth in the low-income countries (Goodland and Daly 1996).

The UNCED process seeks to resolve the impasse created by the South's distrust of the Northern environmental agenda and the North's critique of the socio-political and cultural habits of the South. A starting point is to construct a more accurate and differentiated picture of the Third World. While subsistence livelihoods are a common feature of Third World societies, there are none the less significant divisions and differences within the G-77 countries, as the group of the Third World is known. There are differences in levels of economic development, in the nature of their trade and levels of exports and differences in political stability. In this respect, many international organizations, including the UN, make a distinction between the Third World and least developed countries (LDCs). The UN recognizes fifty LDCs.

Box 7.2 Distinguishing least developed countries

- A *low income* criterion: based on a three-year average estimate of the gross domestic product *per capita* (under US$900).
- A *human resource weakness* criterion: based on indicators of: (1) nutrition, (2) health, (3) education and (4) adult literacy.
- An *economic vulnerability* criterion: based on indicators of: (1) the instability of agricultural production, (2) the instability of exports of goods and services, (3) share of manufacturing and modern services in GDP, (4) merchandise export concentration, and (5) the handicap of economic smallness.

Source: adapted from http://www.un.org/esa/sustdev, accessed 14 February 2004.

Promoting sustainable development in LDCs is particularly problematic. They are among the most vulnerable in the international trading and economic system and have the least capacity to respond to global environmental change, including climate change, although these changes will have significant adverse effects for many of them. Caught up in a web of international indebtedness, their capacity for environmental management is also weak, as countries have neither the resources nor the expertise to begin to manage their environment in ways that are relevant to local long-term needs.

Recognition has also to be given to the fact that there is not 'one set of villains and another of victims' (WCED 1987: 47). Social inequalities also exist within the Third World. Many Southern countries have political and economic elites who enjoy affluent, Western *lifestyles*. In contrast, the bulk of their populations, especially in rural areas, are struggling to eke out a *livelihood*. It is also important not to construct an image of Third World communities as passive victims of wider global processes. The recent growth of resistance movements against globalized agriculture and biotechnology and the anti-globalization alliances that were witnessed in action in Seattle in 1999 provide ample counter-evidence to such a view.

The UNCED process draws upon several tools to help the Third World in the implementation of the agreements, principles and conventions reached at the UN Summits (Box 7.3). Implementation efforts occur through the myriad of international organizations and financial regimes that are involved in shaping the economic, environmental and development strategies of Third World countries. One way of exploring these regimes and their impact on the prospects for, and barriers to, the promotion of sustainable development is to examine key themes that are emerging in the sustainable development agenda of the Third World. These key themes are used as lenses through which to explore whether, and in what ways, the global governance regime promoted by UNCED is responsive to the needs, values and aspirations of the Third World.

Box 7.3 Range of implementation tools

Introducing good governance practices

- Building capacity at national and international levels.
- Developing democratic and accountable institutions.
- Enforcing international legal agreements.
- Promoting good governance practices within and between states.
- Ensuring the involvement of major groups and stakeholders.
- Developing new partnerships (for example, Type II partnerships).
- Utilizing science and knowledge for sustainable development.
- Providing education and training and increasing public awareness.
- Providing and sharing information.
- Monitoring progress and identifying implementation barriers.

Relating sustainable development to trade and globalization

- Reducing or eliminating tariffs on non-agricultural and agricultural products.
- Transferring environmentally sound technology.

Providing financial assistance

- Strengthening financial resources and mechanisms.
- Increasing financial transfers from multilateral institutions, such as bilateral assistance programmes, GEF and the World Bank.
- Reducing debt.
- Enhancing ODA transfers from donor countries.

The sustainable development agenda: themes from the Third World

This section explores five key themes that link environment and development, in the context of the promotion of sustainable development, in the Third World.

Theme 1 Setting a relevant agenda

Third World countries have a range of sustainable development priorities, which reflect their social, political and economic contexts. Many of these priorities stand in sharp contrast to the priorities of the industrialized world, leading to concern that, for example, the Northern environmental agenda does not necessarily reflect the principal environmental problems faced by the Third World. From the list in Box 7.4 it can be seen that the promotion of sustainable development in the Third World has to be undertaken in parallel with action directed at realizing several interrelated development goals. In this context, Third World countries have to overcome multiple, and very pressing, obstacles (Box 7.5).

Box 7.4 Key sustainable development priorities for the Third World

Economic development

- Eradicating poverty.
- Supporting sustainable agriculture and rural development.
- Ensuring fair wages, health, and safety in the workplace.

Social development

- Overcoming illiteracy and improving access to education.
- Improving the position of women.
- Providing sanitation and safe drinking water.
- Making health care accessible and combating disease.
- Building safe and healthy shelter, especially for slum dwellers.

Ecological sustainability

- Upholding sustainable patterns of resource access and use.
- Defending the natural resource base.
- Combating deforestation, desertification and soil erosion.
- Protecting biological diversity.

Box 7.5 Social and political obstacles to sustainable development in the developing world

- Political corruption.
- Armed conflict.
- Trafficking in drugs, arms and persons.

---

continued

- Organized crime.
- Food insecurity.
- Population growth.
- Disease, especially Aids.
- Rapid resource depletion.
- Racial, ethnic and religious tensions.
- Discriminatory attitudes and practices towards women and girls.

While the North also experiences many of these problems, including within the transition states of Eastern Europe, the intensity with which the problems are felt marks Third World societies. The list in Box 7.5 gives an idea of the daunting nature of the tasks ahead and points to how the promotion of sustainable development is linked, in an integral way, with the resolution of complex social, economic and political problems. Yet the list reveals nothing about the *causes* of these problems. Why is the Third World so socially, politically, economically and ecologically vulnerable? As it discusses further the sustainable development themes that have emerged in Third World discourses and practice this chapter points to some answers to that question.

Theme 2 Gender and the environment

Governance networks

The linkages between gender and the environment form another common theme in Third World sustainable development discourse, although this theme has also major relevance to the industrialized world (Buckingham-Hatfield 2000). That the Brundtland conceptualization of sustainable development and the global efforts of UNCED now recognize a gender dimension is largely due to the activities of women's groups. In particular, it owes much to the activities during the UN International Decade of Women (1976–85) and the work of the UN International Research and Training Institute for the Advancement of Women.

During these early years the so-called 'women, environment and development' (WED) debate framed this discussion. Women's networks participated in the 1992 Rio Earth Summit. The policy document *Women's Action Agenda 21* (see Box 7.6) and the Planeta Femea event held at the Global Forum at Rio helped to infuse a gender perspective into the output of the Rio Summit. Because of these activities, gender is now an established item on the international environment and development agenda.

Box 7.6 Governance networks: the WEDO example

In 1990 the Women's Environment and Development Organization (WEDO) was established. It acted as a lead voice for women in the run-up to the Rio Earth Summit. Its policy document *Women's Action Agenda 21* served as a basis for introducing sections on gender equality in Agenda 21 and the Rio Declaration. WEDO has since participated in all the UNCED summits, as well as the major international UN conferences on development, including those at Beijing, Istanbul and Cairo.

The mission of WEDO is 'to increase the power of women worldwide as policy makers in governance and in policymaking institutions, forum and processes, at all levels, to achieve economic and social justice, a peaceful and healthy planet and human rights for all'.

WEDO has helped to establish the Women's Caucus, which acts as an advocacy group, advancing women's perspectives at the UN and other inter-governmental fora.

The related *Women's Action Agenda for a Healthy and Peaceful Planet 2015* served as a basis for women's lobbying during the WSSD events.

Source: adapted from http://www.wedo.org/about/about.htm, accessed 19 January 2004.

Gender and the conceptualization of sustainable development

The WED discourse highlighted the links between the social and economic position of women in the Third World and environmental degradation. The position of women makes them more vulnerable to the negative effects of environmental degradation than their male counterparts. They are more marginalized, usually work harder, especially if engaged in agricultural labour, have a less adequate diet and are often denied a voice in the political, economic and social spheres. The WED discourse helped focus attention on how the accelerated environmental degradation of the South made women's daily search for firewood, fodder and water more difficult. The discourse made an explicit link between gender inequality and the impact of environment degradation. The controversial link between population size and environmental degradation also came to the fore.

However, the early years of the WED debate presented women as passive victims of the environmental degradation stemming from global processes. As the discourse shifted from discussion of the environment to that of sustainable development, a new focus emerged. This emphasized women's positive role as efficient environmental resource managers within the development process in the South.

Box 7.7 Gender and sustainable development: the links

Gender-specific impact of environmental degradation

- Women make up the majority of the world's poor. Poverty is caused by, but also contributes to, environmentally harmful patterns of natural resource use.
- Given their economic and social roles, environmental degradation has a disproportionate impact upon the daily lives of women.
- Given their vulnerable economic and social positions, the negative aspects of globalization (growing inequalities, inequitable distribution of wealth and resources) have a disproportionate gender impact.
- Environmental security and health issues also have a gender dimension: given women's reproductive and social roles, environmental hazards have different effects on the lives of women and men.

Women and the promotion of sustainable development

- Because of their domestic, agricultural and cultural roles, women are key agents in the promotion of sustainable patterns of natural resource management.
- Women are holders of knowledge about their local environment (indigenous ecology). This stems from their role in the provision of food and traditional medicine, and they are therefore the key to developing appropriate biodiversity preservation strategies.
- Promoting sustainable livelihoods at the community level can be accelerated by giving women the right to inherit land and to have access to resources and credit.
- There is a strong connection between promoting human rights, especially for women, and promoting sustainable development, which is based upon principles of equity and partnership.
- Promoting democratic environmental governance has a gender dimension: participation based on the principle of gender equality of access is more legitimate, democratic and effective.

There are many examples of women's activities that have helped to promote a gender-sensitive sustainable development trajectory. These involve campaigns to protect traditional ways of life, reverse ecological damage and undertake ecological regeneration projects. India's Chipko, or tree hugging movement, formed in the 1970s, is among the most famous of these movements. It helped developed a 'feminist forest paradigm' that has been influential across the Himalayas and beyond (Shiva 1989). Similarly, the crisis in women's access to wood and water motivated the Nobel laureate Wangari Maathai to launch the Greenbelt Movement (Maathai 2003). The Environmental Movement of Nicaragua explicitly deals with women's issues, in particular the high levels of exposure of women agricultural workers to pesticides (Box 7.8).

Box 7.8 The links between women, environment and health: awareness of the dangers of pesticides

UNEP estimates accidental poisoning from exposure to pesticides causes 20,000 deaths and 1 million illnesses worldwide every year. To understand the gender implications it is helpful to consider:

- the differential use of pesticide by men and women during agricultural production;
- the unique health effects on women;
- the extent of information about pesticides available to each gender.

Women farmers and workers are frequently directly exposed to dangerous pesticides. Impacts on women's reproductive health include a greater incidence of miscarriages and stillbirths and an increased incidence of birth defects. There are also potential carcinogenic effects. An extreme example is DDT, once widely used for controlling insect pests on agricultural crops. DDT is highly persistent in the natural environment and accumulates through the food chain. It increases the risk of breast cancer, and an infant feeding on breast milk can receive up to twelve times the acceptable limit of DDT. DDT is now illegal in many countries but it is used in many Third World countries, as it is cheaper than less persistent alternatives.

Chapter 14 of Agenda 21 recommends increased awareness of sustainable agricultural methods in women's groups. These include reducing the use of agricultural chemicals as well as making wider use of traditional practices for pest control. Examples of gender-sensitive activity in this area include 'Field Schools' in Asia, the Environment and Development Action Network in Africa and 'Mama 86' in Ukraine.

Source: adapted from http://www.wedo.org/about/about.htm, accessed 19 January 2004.

Eco-feminism

The argument for the increased participation of women in environmental management is built upon a claim that women had 'privileged knowledge and experience of working closely with the environment' (Braidotti *et al.* 1994: 2). However, some have gone further than this, stressing that women have a special relationship with nature, a claim known as eco-feminism (Box 7.9).

Box 7.9 Eco-feminism

Eco-feminism is both an analysis of society–nature relations and a prescription of how these relationships can be transformed (Buckingham-Hatfield 2000).

continued

Its analysis is highly critical of the principal philosophical and cultural attitudes that underlie mainstream Western ideologies about women, the natural world and their interrelationship (Baker 2004). Eco-feminism draws upon the feminist theory of patriarchy and combines it with insights gained from environmental and peace activism. A central argument is that a common dualistic belief system, rooted in the principle of domination and subjugation, underlies modern, negative attitudes to both women and nature (Plumwood 1986).

To counteract this dominant ideology, eco-feminism aims to reconstruct a new understanding of the place of human beings within the natural world. In particular, it aims to situate women, nature and, sometimes, men in a more balanced and equitable relationship with each other (Diamond and Orenstein 1990).

Eco-feminism, as political activism, arose from what had hitherto been two different social moments, the environmental movement and the women's movement. From within the latter it has inherited much from the women's peace and spirituality movements. It seeks to counteract the myriad ways in which degradation of the natural environment influences the daily lives of women, especially in the Third World.

A convention has grown up in the literature that divides eco-feminism into two broad groupings: 'cultural eco-feminism' and 'socialist eco-feminism'.

Cultural eco-feminism draws heavily upon the tradition of radical feminism (Spretnak 1990). Radical feminist analysis has located women's oppression with men, particularly with male sexuality, regarded as the site of male power. Central to radical feminism is the belief that there exists an innate female nature, which differs from the gendered self of the male. Cultural eco-feminists have extended this central tenet, to argue that women, by virtue of their biological capacity, have a closer relationship with the natural world than do men. This is significant for social movement activity: it has allowed women to keep sight, throughout history, of the mutually interdependent relationship that exists between humanity and the natural world (Merchant 1980). Thus women are in a unique and advantageous position to engage politically, culturally and socially on behalf, and in defence, of nature. This 'standpoint theory', as it is known, is based on the belief that only those who are oppressed can understand and counteract the relationship of oppression.

The claim that women have an innate nature or essentialist characteristics is called *essentialism*. The return to essentialism is the source of considerable disquiet, one could almost say alarm, among the broader, contemporary, feminist movement. Many contemporary feminists strongly distrust the eco-feminist reconnection of femaleness with the sphere of reproduction and with nature. Such connections are dangerously conservative, representing the antithesis of the aims of the post-war Women's Liberation movement.

Acceptance or rejection of essentialism provides a key way of distinguishing the two main tenets of eco-feminism. It is in its efforts to bypass the morass of essentialism that the second main type of eco-feminism, materialist, socialist eco-feminism (at times also referred to as socialist–anarchist eco-feminism) comes to the fore. Crucial to the socialist eco-feminist position is the claim that the exploitation of nature relates to exploitation in society. The reason why women have different experiences of nature from men is due not to their 'essential nature' but, rather, to the fact that we live in a gendered society (Pepper 1996). Women do not have a natural affinity with nature; rather, the link between women and the environment is socially and culturally constructed (Agarwal 1992).

Source: adapted from Baker (2004).

Some feminists have severely criticized the argument that women should provide the moral and the practical efforts necessary to reverse environmental deterioration. While promoting sustainable development can be seen as progressive, 'cleaning up' after men is not and conforms to existing stereotypes about women and their role in society. Women's involvement in the provision of Primary Environmental Care, for example, as promoted by some Third World agencies, is seen as adding to women's daily burden. While participatory and community-based, it none the less equates 'community' work with voluntary, unpaid work by women. This form of Green participation is not ultimately liberating for women (Agarwal 1997).

Box 7.10 Gender and the three pillars of sustainable development

Environmental protection

- Requires understanding of the gender-specific impacts of environmental degradation and misuse.
- Requires recognition of women's relationship to environmental resources and their roles in resource planning and management.
- Requires incorporation of women's knowledge of environmental matters into policy and planning.

Economic well-being

- Requires gender-sensitive strategies: 70 per cent of the world's estimated 1.3 billion people living in absolute poverty are women.
- Requires recognition that the economic well-being of any society cannot be achieved if one group is massively underprivileged compared with the other.
- Requires realization that an economy cannot be called healthy without utilizing the contributions and skills of all members of society.

continued

> **Social equity**
>
> - Requires making the link between gender equity and social equity.
> - Requires acceptance that no society can survive in the long run, or allow its members to live in dignity, if there is prejudice and discrimination against any social group.
>
> *Source:* adapted from Hemmati and Gardiner (2001).

Theme 3 The environment, trade and the WTO

Another theme in the relationship between development and the environment in the Third World arises from the strong links between trade liberalization policies, or free trade, and the environment. In 1994 the major international organization that promotes trade liberalization, the World Trade Organization (WTO), was established. The WTO is also responsible for enforcing international free-trade law. Its mission is to promote:

> the optimal use of the world's resources in accordance with the objective of sustainable development. . . . [C]ontributing to these objectives by entering into reciprocal and mutually advantageous arrangements directed to the substantial reduction of tariff and other barriers to trade and to the elimination of discriminatory treatment in international trade relations.
>
> (GATT 1994: 9)

The free-trade system upheld by the WTO is an important linchpin in the global economic system. It upholds traditional economic values, particularly with respect to the view that free trade can be a route to modernity, especially for the Third World. Free trade, it is argued, encourages a country to specialize according to its comparative advantage and, by exposing countries to competition, also forces more efficient resource use. In addition, an environmentally regulated free-trade system can apply sanctions to countries with low environmental standards.

Despite these claims, many argue that trade regulations restrict the ability of states, particularly those in the Third World, to promote sustainable development. It is not surprising, therefore, to learn that the WTO has received considerable attention from environmental activists, especially from anti-globalization protesters. Critics point to the lack of transparency in the way the WTO conducts its business. Trade experts dominate meetings, which remain closed to environmental and civil society organizations. This behaviour is not in keeping with the principles of good governance. However, the direct relationship between environmental protection measures and free-trade policies is also a source of more conflict. To deal with the conflict, the WTO has established a Committee on Trade and the Environment.

One of the issues that the committee addresses is the use of trade restriction measures in multilateral environmental agreements (MEAs). Among the 200 MEAs already in existence, the WTO estimates that twenty include environmental trade measures. The 1973 Convention on the International Trade in Endangered Species of Wild Flora and Fauna (CITES), the 1987 Montreal Protocol on Substances that Deplete the Ozone Layer and the 1989 Basel Convention on the Transboundary Movement of Hazardous Wastes and their Disposal are all good examples of MEAs that restrict trade.

A second area of concern is the trade disputes that arise when environmentally based distinctions are made between what would otherwise be considered 'like' goods, such as distinguishing 'dolphin-safe tuna', fur from animals not caught in leg-hold traps and beef produced without artificial hormones. The WTO abhors such distinctions, seeing them as a guise for protectionism. In contrast, such distinctions are seen by many as the backbone of environmental policy (Moltke 1997). Thus environmentalists reacted with alarm when the WTO ruled that the US could not ban shrimp caught with nets that trap turtles (Dresner 2002), a ruling that has become notorious among environmental activists.

Another area of disquiet is the impact of WTO rulings and agreements on national environmental protection measures. When national governments take such measures, they can lead to restrictions on the importation of goods that do not conform to certain environmental standards. The WTO sees this as a restriction on free trade. National actors see such measures as a way in which they can uphold their environmental standards. There is also a fear that trade liberalization pro-motes a race to the bottom, as countries come under competitive pressure to lower their environmental standards. The conflicts between free trade and environmental protection have come into ever sharper focus as the pressures of globalization, and the related increase in international trade, continue.

Increasingly, there are calls for the development of a trade regime that enhances, not destroys, the prospects of promoting sustainable development. A meeting of the WTO in Doha, Qatar, in 2001 led to the Doha Declaration, which contains both a declaration outlining the beliefs and commitments of the WTO and a Work Programme to put the commitments into effect. The Doha Declaration (Box 7.11) is a response to the growing opposition to the world trade regime, as witnessed at the riots in Seattle in 1999, while at the same time it reiterates the WTO belief in the value of free trade for the promotion of sustainable development.

Box 7.11 The Doha Declaration: key points

Ministerial declaration

- The multilateral trading system embodied in the WTO has contributed significantly to economic growth, development and employment.
- International trade can play a major role in the promotion of economic development and the alleviation of poverty.
- Least developed countries are particularly vulnerable and face special structural difficulties in the global economy.
- The challenges that members face in a rapidly changing international environment cannot be addressed through measures taken in the trade field alone.
- The aims of upholding and safeguarding an open and non-discriminatory multilateral trading system, and acting for the protection of the environment and the promotion of sustainable development, can and must be mutually supportive.
- Under WTO rules, no country should be prevented from taking measures for the protection of human, animal or plant life or health, or of the environment, subject to the requirement that they would not constitute a disguised restriction on international trade.

Work programme: aims

- *Agriculture*: to improve market access; to bring about reductions of, with a view to phasing out, all forms of export subsidies; to achieve substantial reductions in trade-distorting domestic support; to support differential treatment for the Third World, to enable it to take account of their development needs, including food security and rural development.
- *Market access for non-agricultural products*: to reduce or eliminate tariffs, as well as non-tariff barriers, in particular on products of export interest to the Third World.
- *Trade-related aspects of intellectual property rights*: to implement the Agreement on Trade-related Aspects of Intellectual Property Rights (the TRIPS Agreement) in a manner supportive of public health, by promoting both access to existing medicines and research and development into new medicines. The relationship between the TRIPS Agreement and the Convention on Biological Diversity and the protection of traditional knowledge and folklore needs to be examined.
- *Relationship between trade and investment*: developing and least developed countries need enhanced support for technical assistance and capacity building.
- *Trade and the environment*: to begin new negotiations on the relationship between existing WTO rules and specific trade obligations set out in MEAs; to support the reduction or, as appropriate, elimination of tariff and non-tariff barriers to environmental goods and services; to give attention to the effect of environmental measures on market access, especially for the Third World.

- *Trade, debt and finance*: to examine the relationship between trade, debt and finance to contribute to a durable solution to the problem of external indebtedness of developing and least developed countries.
- *Least developed countries*: to support the concerns expressed by the LDCs; to recognize that to integrate the LDCs into the multilateral trading system requires market access, support for the diversification of their production and export base, and trade-related technical assistance and capacity building.

Source: adapted from WTO (2001).

The Doha Declaration acknowledges that the promotion of sustainable development is a collective responsibility, and it accepts the need for differential treatment. It also directs attention to matters relating to trade and intellectual property rights (TRIPS). In addition, it places the onus on the WTO to address some of the underlying obstacles to sustainable development in the Third World and to support technical assistance and capacity building. However, the WTO is hoping that an enhanced programme of technology transfer and capacity building, designed to deal with the problem of weak trade capacity, will resolve the issue. Despite these limitations, the declaration is important, in that it recognizes the expanded nature of the agenda of sustainable development, which now encompasses issues of environment, development *and trade*.

However, there was hope that the Doha meeting would do more than this and, in particular, that it would clarify the relation between MEAs and the WTO regime. It was also hoped that the meeting would push sustainable development to the top of the WTO agenda, making it an overarching goal (Tunney 2004). The declaration, and its related Work Programme, can also be criticized for their limited understanding of the difficulties faced by the Third World in the global economy. While recognizing the need to deal with debt, it none the less reduced the problems of the Third World to those of a technical nature, such as the lack of technical and administrative expertise, which are seen to limit their capacity to realize the benefits of international trade. The solution is to support an international programme of enhanced transfer – so that best practice, technology and financial resources can be transferred from the industrialized world.

Critics have argued that this ignores the possibility that such transfers may increase the structures of dependence that tie the Third World into an inequitable relationship with the industrialized world. In addition, the underlying premise of the declaration is that trade and the promotion of sustainable development can be mutually supportive. However, Green theorists and supporters of strong sustainable development criticize this assumption. They argue that free trade, the development of global markets and the stimulation of economic growth are not consistent with the principles of localism and equity, essential for the promotion of sustainable development (Pepper 1996). In addition, free trade is seen as having resulted in a devastating pattern of environmental asset stripping, to the benefit of the industrialized world and to the detriment of the developing world.

> The use of 'free trade' to cement global economic integration removes the principle of sustainability from local communities across the globe. It does so by stimulating the trade in natural resources, and their products, without strengthening local communities or encouraging responsible environmental management. Resources are depleted to provide foreign exchange, and sustainable livelihoods are eroded.
>
> (Redclift 1997: 395)

Theme 4 Knowledge, science and sustainable development policy

The relationship between knowledge, science and policy forms a growingly important theme. Chapter 35 of Agenda 21 addressed the relationship between science, knowledge and the promotion of sustainable development (Box 7.12).

The UNCED process has given a new role to experts and their scientific knowledge, both in identifying and in providing solutions to environmental problems. Some would argue that this has enabled science to play a *dominant* role in setting the priorities of the international environmental agenda (Blowers 1997). This has led to controversies, including, for example, in relation to the role of science in the development of new commercial uses for plant and animal genetic resources (Box 7.13).

Box 7.12 Chapter 35 of Agenda 21

- Scientific research should support the search for appropriate strategies for sustainable development.
- The scientific base should be widened to meet the needs of environmental and development management.
- The natural environment and changes caused by human beings are interlinked, thus our scientific understanding should be enhanced.
- Long-term scientific assessment of present and future situations is needed. A standard methodology should be developed.
- Scientific capacities on environment- and development-related issues should be promoted, particularly in the Third World.
- Institutional partnerships and multidisciplinary activities, including on policy strategies, should be promoted.

Source: adapted from Koch and Grubb (1993).

Box 7.13 Science, biotechnology and indigenous knowledge

The debate over biotechnology provides a good example of disputes over the relationship between science, economy and ecology. On the one hand, the biotechnological revolution is presented as an answer to the loss of biodiversity, caused by intensive agriculture and deforestation. By giving a use value to plant and animal genetic resources, biotechnology provides a rationale for halting the destruction of the planet's rich biodiversity. These resources can then be put to use to develop new medicines and products.

In the contrasting view, the application of science is seen as an example of on-going colonial relations between the North and the South, especially when account is taken of the fact that two-thirds of the planet's species are in the Third World. One particularly contentious issue is that the development of the biotechnology industry has led to the creation of intellectual property rights in relation to natural resources – for example, patents on seeds. The knowledge of pharmaceutical companies is protected by international law through the patent system. However, international law sees the knowledge of indigenous peoples – for example, their knowledge of the medicinal properties of plants – as traditional and not 'novel', and therefore it can be obtained without payment. This is seen to serve the interests of corporations at the expense of peasants' and farmers' rights and to fail to guarantee the long-term survival of species. It privileges individuals, states and corporations over indigenous peoples or local communities. This has led to a growing controversy over Trade-related Aspects of Intellectual Property Rights (TRIPS). Such a system also legitimizes private intellectual property rights over life forms, which many find ethically contentious.

Source: adapted from Banerjee (2003).

The controversies over biotechnology help to show that knowledge is embedded in wider systems and practices that are not neutral. The application of science can also have social and economic effects, as shown by the way in which distortions in the granting of intellectual property rights can marginalize the knowledge, practices and traditional economies of many Third World communities (Shiva 2000).

There are other elements to the debate. While scientific knowledge is presented as authoritative, it is more often uncertain and contestable (Yearley 1997). It can make mistakes, it can bequeath new environmental problems to future generations – for example, nuclear waste – and it can result in the reduction of biodiversity, as has happened with the development of scientific agriculture. Scientific agriculture has led to modern practices of mono-crop production in many Third World countries, such as coffee or banana plantations. This form of intensive farming is dependent upon the application of scientific methods of pest control and artificial soil enhancement. Intensive agriculture has created environmental problems that, in turn, require the further application of scientific solutions, leading to a spiral of dependence.

There is also concern that the knowledge used to construct global environmental management regimes draws upon a narrow, Western approach and not the indigenous ecology that has been used to 'manage' local environments over an extended period of time (Redclift and Woodgate 1997). Scientific ecology, for example, has played a role in the construction of Western models of best practice for nature conservation, which have encouraged the designation of vast tracts of land as 'nature reserves'. There are several documented cases where this has displaced the communities that have depended on such land for subsistence, while opening the reserves up for the pleasure of Western tourists. This happened in the designation of the land of the Chenchu community in southern India as a tiger reserve.

Theme 5 Reconstructing global institutions of financial governance

The World Trade Organization, the International Monetary Fund (IMF) and the World Bank are powerful agents in advancing particular discourses as well as practices in relation to implementing sustainable development policies in the Third World. Focusing on this theme helps to turn attention to the role of international institutions in the governance of sustainable development.

Global Environment Facility

The Global Environment Facility (GEF) was established in 1991 as a cooperative venture between the United Nations Environment Programme (UNEP), the United Nations Development Programme (UNDP) and the World Bank, in the run-up to the Rio Earth Summit. The GEF is an important part of the institutional structures established under the Rio process. It is the primary funding mechanism for dealing with issues of the global environment; it forms the linchpin in the financial deal reached between the industrialized countries and the Third World at the Rio Summit; and it also acts as the financial mechanism for both the UNFCCC and the CBD. Its mandate is to fund the costs to the Third World of its efforts to limit the global impacts in four key environmental areas: ocean pollution, climate change, biodiversity loss and ozone depletion. However, it can fund only the incremental costs of meeting the Conventions agreed at Rio.

In practice, funding the promotion of sustainable development has proved to be both a difficult and a contentious problem. Despite intense and protracted discussions at Rio and the expectations of the Third World, there is no satisfactory financial arrangement for funding the measures agreed in Agenda 21. At Rio, the North agreed to provide 'new and additional' financial resources to the Third World to enable it to meet the costs involved in implementing the agreements reached. Chapter 33 of Agenda 21 reiterated the target of industrialized countries transferring 0.7 per cent of GNP as aid to the poorest countries. However, five years after the Rio Summit, the UNGASS New York Summit came close to collapse on the issue of North–South finance. In the period 1991–5 total levels of ODA fell. In the period 1991–6 the GEF budget was US$3 billion, but ODA fell by US$8 billion during the same period (Lake 1998). In addition, the US is in arrears in meeting its financial commitments to the GEF.

As a result, there were fears that the WSSD would be embroiled in acrimonious discussions over finance, which would doom the Johannesburg Summit to failure. Consequently, there was a strategic decision to get heads of state and government to attend an International Conference on Finance for Development in Monterrey, Mexico, in 2002. As a result only very limited discussion took place at Johannesburg about financing the promotion of sustainable development. The Monterrey meeting resulted in the so-called Monterrey Consensus, which agreed the following goals: 'to eradicate poverty, achieve sustainable economic growth and promote sustainable development as we advance to a fully inclusive and equitable economic system' (Box 7.14).

Box 7.14 Monterrey Consensus

The aim is to confront the challenges of financing development, particularly in the Third World. The goal is to eradicate poverty, achieve *sustained economic growth and promote sustainable development*, to advance to a fully inclusive and equitable global economic system. Concern is expressed about current estimates of dramatic shortfalls in resources required to achieve the internationally agreed development goals, including those of the UN Millennium Declaration.

Actions

- Mobilizing domestic financial resources for development.
- Mobilizing international resources for development, particularly through foreign direct investment.
- Using international trade as an engine of development.
- Increasing international financial and technical cooperation for development.

---

continued

- Dealing with external debt and providing debt relief measures.
- Addressing systemic issues: enhancing the coherence and consistency of the international monetary, financial and trading systems in support of development.
- Staying engaged through building a global alliance for development.

Source: adapted from UN (2002a).

In Monterrey the US and the EU committed a total of US$ 30 billion, subject to the implementation of good governance principles in the beneficiary countries. Monterrey also provided an impetus for the private sector to take responsibility for promoting sustainable development.

Subsequently, the Johannesburg Summit agreed a replenishment of the GEP by US$3 billion. Meeting financial commitments is just one problem. There are also institutional difficulties. The World Bank houses and manages the GEF secretariat and the GEF Trust Fund. The central role that the World Bank plays keeps the GEF under critical and constant scrutiny (Elliott 2002). From the perspective of the Third World and many environmental and developmental NGOs, the relationship between the GEF and the World Bank is too close. Despite the restructuring of the GEF in 1994, concern over the transparency and account-ability of its governance structure remains. As a result, there is an uneasy relationship between the GEF and the Third World, especially noticeable in the relationship between the CoP of the CBD and the GEF, although both signed a Memorandum of Understanding in 1996.

The World Bank

The post-war period brought the countries of Asia, Latin America and Africa into the international system of global macroeconomic management through the work of the Bretton Woods institutions (the International Monetary Fund and the World Bank). A free-trade ideology drives these institutions. The lending programme of the World Bank, designed to enable the Third World to 'catch up' with development, resulted in a spiral of debt and poverty. In the 1980s the majority of the Third World was caught in a cycle of debt when the interest rate on the so-called 'cheap' loans taken out in the 1970s rose dramatically. Many Third World countries were forced to call in the International Monetary Fund to avoid complete bankruptcy. The IMF imposed tough 'structural adjustment' conditions on the new round of loans, including drastic cuts in public expenditure, especially on welfare and education. In addition, countries were put under pressure to increase their exports to pay for the debt, a process that encouraged unsustainable exploitation, especially of forest resources (Reed 1996, 1997; Dresner 2002). These programmes also had gender-specific impacts (Baker 1994).

Given the historical connection between structural adjustment policies and World Bank lending, the Bank has become subject to 'relentless' attention from NGOs and grass-roots organizations. Critics point to the environmental impact of projects funded by the Bank, especially infrastructure projects, such as large dams, that have resulted in the dislocation of local, often tribal, people. While it has become institutionally crucial to the promotion of sustainable development, it has, in the words of one critic, 'become materially central to continued environmental decline in those countries in which it funds projects and programmes' (Elliott 2002: 65). The World Bank has also become the target of environmental NGOs in the hope that, by focusing on the source of environmental failures in the development process, they can influence the policies of developing countries themselves (Reed 1997).

Having been subjected to severe criticism, the World Bank has undertaken several changes, such that many now argue that it is a leader among multilateral institutions regarding environmental standards of behaviour. The Bank has an Environment Department and a Vice-Presidency for Environmentally Sustainable Development, created in 1993. This focuses on integrating economic, ecological and social considerations into the projects it funds. It is most widely known for having commissioned various reports on environmental assessment. The World Bank obliges borrowing governments to meeting stringent environmental standards (Box 7.15). It is fair to say that these reforms are more ambitious and inclusive than the institutional changes implemented by any other multilateral development bank (Reed 1997). They have enabled the World Bank to set the pace and standard for other international organizations whose behaviour has environmental consequences.

Box 7.15 Environmental standards of the World Bank

- Encourages governments to develop national environmental action plans as the basis of lending operations to all economic sectors.
- Expanded the number of projects subject to environmental assessments.
- Revised the composition of its lending portfolio, with increases in cumulative funding for environmental projects.
- Holds that people should be resettled only if the displacees' prior income levels will be restored.
- Developed an information policy that makes information available about approved projects, and has established public information centres.

continued

- Established a formal review process, through which the public can submit complaints.
- Established twenty operational directives that address environmental issues, ranging from specific resource concerns, such as forests and water, to those that concern the impact on indigenous peoples.
- Expanded the number of resource-based projects, as well as the number of project loans that have natural resource components, while changing its approach to the use of natural resources.
- Helped to establish the GEF.

However, there is still a disjuncture between the formal requirements set by the Bank and actual respect for those standards in the design and implementation of projects. This is very often because many Third World countries lack the will or the ability to respect the standards in practice (Reed 1997). More fundamentally, there is the argument that the World Bank helps to accelerate the integration of the Third World into the dominant growth-orientated development paradigm. This condemns the Third World to a subordinate position in the international economic order. It also represents an example of what *The Ecologist* has referred to as the 'closure of the commons' – that is, bringing more and more areas, countries and resources under the remit of the international economic order.

There is also the criticism that the World Bank imposes a Western-centric idea of development. Part of the task of those who seek to promote sustainable development is to construct a new understanding of development, one that recognizes the diversity of development paths that are needed in order to take account of the cultural, economic and ecological context within which development takes place. Critics also hold that the World Bank has not accepted the premise that humankind has inherited limited environmental stocks, which imposes irreducible constraints on the economic system. It therefore continues to believe that economic growth need not be constrained by 'external' factors. Rather, the substitutability of human-made for natural capital, coupled with technological innovation, will enable humankind to overcome any environmental constraints. This is a weak sustainable development position. Technological transfer and capacity building thus become the key ways to resolve environmental problems and so promote sustainable development. This 'allows the human community to cling to unrealistic expectations regarding achievable standards of living for the great majority of humanity and to believe that global inequalities and poverty can be addressed by more growth in both the North and the South' (Reed 1997). From this perspective, the World Bank forms part of the structural causes of unsustainability.

Conclusion

In this chapter the promotion of sustainable development in the Third World has been shown to be a pervasive task, that raises multiple, complex and highly contested issues. Efforts are constrained by an international economic and political system that is orientated to the promotion of values, norms and principles, such as the principle of free trade, which are incompatible with the principles of sustainable development. The trade system promoted by the WTO is not easily reorientated towards a sustainable development agenda.

Exploration of the ways in which structural adjustment policies, the burden of external debt and the liberalization of trade have contributed to unsustainable development patterns in the South has pointed to a recurring theme in this book: the need to address the inequitable basis of the global economic and political system. Promoting sustainable development requires breaking the causal connections between environmental degradation, poverty and population growth. Not only does this require a reduction in consumption levels in the high-consumption societies of the West, but ultimately it calls for a fundamental restructuring of the system of international political economy.

Adopting a Third World perspective, some critics are highly sceptical of the discourse on, and engagement with, sustainable development that have emerged since the Rio Earth Summit. The UNCED process assumes that the resolution of global environmental problems requires large-scale capacity transfers to the poorer countries. This would enable the South to leapfrog the North's environmentally damaging stage of economic evolution. Critics argue that the values that this approach promotes are so overwhelmingly Western in origin and interest orientation that they undermine the equity and justice agenda that the environmental crisis calls for. They point out that the Western project to modernize post-colonial societies has contributed to poverty, to increases in economic and gender inequality and to environmental degradation, which, in turn, further diminishes the life chances of the poor.

However, taking a Brundtland perspective, as opposed to limiting attention to what has been achieved through UNCED, a different picture emerges. While the UNCED process stresses the importance of technological transfer, Brundtland recognizes the need for a more fundamental approach, that requires a reduction in resource use in the North, linked with changing consumption patterns. Ultimately, promoting sustainable development requires overcoming the causes of unsustainable development that stem from the inequity in the global economic system. While the sustainable development agenda of international institutions such as the World Bank is flawed, the Brundtland formulation still contains powerful arguments for radical change.

Summary points

- The environmental crisis in the South is also a warning signal about the development model in the North.
- Constructing a new development paradigm for the Third World that is ecologically and socially aware and sustainable in the long term is premised upon addressing this complex range of issues.
- The promotion of sustainable development in the Third World hinges on building policies and processes that confront five key themes: (1) setting a relevant policy agenda; (2) dealing with the gender-specific dimensions; (3) recognizing the negative relationship between the current global free-trade regime and the promotion of sustainable development; (4) admitting the power base of science, while acknowledging the validity of different types of knowledge, as well as recognizing that indigenous knowledge sources can have a role in the promotion of a sustainable future; (5) reforming the institutions that finance the promotion of sustainable development.
- North–South confrontation on the environment and sustainable development pivots around two axes: (1) poverty, population and consumption; (2) the inequitable nature of the global economic system. The first view seeks to promote sustainable development through technology transfer, capacity enhancement and the transfer of funds. It also calls for more efficient use of resources in the North and, equally controversially, a reduction in population growth in the South. The second view calls for a fundamental restructuring of the international political and economic order.
- Those who present a more radical analysis of, and seek a more fundamental solution to, the problem of global inequality reject the extension of the sustainable development paradigm to the Third World.
- However, the Brundtland model of sustainable development contains a radical agenda of social transformation.

Further reading

The relationship between the North and the South

Escobar, A. (1995) *Encountering Development: The Making and Unmaking of the Third World, 1945–1992*, Princeton, NJ: Princeton University Press.

Esteva, G. (1992) 'Development', in W. Sachs (ed.) *The Development Dictionary*, London: Zed Books, 6–25.

Lefebvre, H. (1991) *The Production of Space*, Oxford: Blackwell.

Redclift, M. (1987) *Sustainable Development: Exploring the Contradictions*, London: Methuen.

Said, E. (1993) *Culture and Imperialism*, London: Vintage Books.

Knowledge, power and the environment

Shiva, V. (1993) *Monocultures of the Mind: Perspectives on Biodiversity and Biotechnology*, London: Zed Books.

The gender perspective

Agarwal, B. (1992) 'The gender and environment debate: lessons from India', *Feminist Studies*, 18: 119–58.

Braidotti, R., Charkiewicz, E., Häusler, S. and Wieringa, S. (1994) *Women, the Environment and Sustainable Development: Towards a Theoretical Synthesis*, London: Zed Books.

Buckingham-Hatfield, S. (2000) *Gender and Environment*, London: Routledge.

Shiva, V. (1989) *Staying Alive: Women, Ecology and Development*, London: Zed Books.

Global environmental change

Blowers, A. (1997) 'Environmental policy: ecological modernisation or the risk society', *Urban Studies*, 34: 845–71.

Redclift, M. (1997) 'Development and global environmental change', *Journal of International Development*, 9: 391–401.

The Western perspectives on nature and development

Banerjee, S.B. (2003) 'Who sustains whose development? Sustainable development and the reinvention of nature', *Organization Studies*, 24: 143–80.

Funding and its structures and institutions

Elliott, L. (2002) 'Global environmental governance', in R. Wilkinson and S. Hughes (eds) *Global Governance: Critical Perspectives*, London: Routledge, 57–74.

Reed, D. (1997) 'The environmental legacy of Bretton Woods: the World Bank', in O.R. Young (ed.) *Global Governance: Drawing Insights from the Environmental Experience*, Cambridge, MA: MIT Press, 227–46.

Web sites

http://www.un.org/esa/sustdev for information on UN Department of Economic and Social Affairs, Division for Sustainable Development.

http://www.wedo.org for information on WEDO.

8 Changing times

The countries in transition in Eastern Europe

Key issues

- Diversity of transitions.
- Environmental legacies of communist rule; nuclear safety; new environmental pressures.
- Eastern enlargement of the EU; EU environmental *acquis*; pre-accession funds.
- Environmental policy integration.

This chapter explores the prospects for, and the barriers to, the promotion of sustainable development in the countries in transition in Eastern Europe. The focus is on Central Europe, including the Baltic states, and the Balkans, with particular attention being paid to the new member states of the EU. The enlargement process created an ideal opportunity for the EU to influence the way in which transition states manage their environment, while at the same time putting its own commitment to sustainable development into effect. The chapter focuses on the tensions between the development demands of transition and the promotion of sustainable development. The Eastern enlargement of the EU has created a large geo-political bloc, a powerful trading body and the world's largest internal market. What happens, therefore, within the EU is of global significance. The chapter begins by exploring the nature of transition, the environmental problems of countries in the region and the influence of marketization and democratization. In this context, the promotion of sustainable development is examined by looking at environmental policy integration.

Understanding transition

The communist regimes in Eastern Europe collapsed in 1989 and, since then, countries in the region have undergone a complex process of transition. This involves political democratization and the introduction of market economies. Initially, it was simply assumed that the countries would adopt political and economic models from Western European states and progress to Western economic and political systems. However, it has since become clear that 'transition' is far from a simple linear process (Smith and Pickles 1998). First, while countries in the region adopt new institutions of governance, they *adapt* them to suit their particular country context. An example of this is the variation in the remit, power and degree of autonomy of the environmental protection agencies that have been established in countries in the region. Second, transition is not taking place in a vacuum, but rather involves complex reworking of old social and economic relations, as countries construct new forms of capitalism on, and with, the ruins of their old communist systems. Privatization policy provides a good example: countries sell off their state-owned enterprises, often resulting in complex forms of mixed ownership, which are often unique to the region (Stark 1997).

Transition in Eastern Europe involves the adoption of new institutions of governance and their adaptation to suit the particular country in which they are adopted. This takes place in the context of a complex reworking of old social and economic relations.

There is also a great deal of variation between the countries in the region. First, account has to be taken of the diversity in the cultural, religious and ethnic make-up of the states (and, at times, of sub-state levels). Second, there are differences in their experiences under communism and in relation to the legitimacy of communist rule. For example, levels of industrialization and the degree of centralization of their economies differed, with Hungary and Bulgaria providing contrasting examples. The use of the single term 'Eastern Europe' to refer to the entire region during the period of communist rule masked this political and economic diversity. Third, the nature of the political 'revolutions' that the countries underwent in 1989 varied. The post-communist situation in the Balkans is less stable than that in the Visegrad countries – that is, Poland, Hungary, the Czech Republic and Slovakia. The Balkans have experienced high turnover rates of governments, low public acceptance of change and poor government commitment to economic and political reform, as reflected in the continuing strength of the communist successor parties. The wars that followed the break-up of the former Yugoslavia also contributed to the instability in the Balkan region. Since 1989 these factors have contributed to the emergence of a highly differentiated transition process between the countries of the Balkan region and those elsewhere in Central Europe (Baker 2005b). The legacies of the old regime shape the capacity of the countries to respond to the challenges of the post-1989 period. They remain key factors shaping the prospects for, and barriers to, the promotion of sustainable development in the countries in transition.

Promoting sustainable development in transition societies

Addressing environmental pollution

While retaining large tracts of unspoilt land, often possessing a rich biodiversity, most countries in the region have inherited a heavy legacy of pollution from the period of communist rule (EEA 1999). Indeed, one of the hallmarks of the old Soviet period was its much publicized environmental mistakes and disasters. Environmental problems include poor air and water quality, inadequate treatment and disposal of industrial waste (including hazardous and nuclear waste), soil deterioration and contamination of land. Most countries also lacked a comprehensive waste management strategy and effective legislation. The most notorious area for air pollution is the so-called 'black triangle', covering the Czech and Slovak republics, Poland and the former German Democratic Republic (East Germany). Other areas of especially high concentrations of pollution include the Black Sea and the Danube river basin (EEA 1999; Carter and Turnock 2001).

Other regions of the world, especially Western Europe, also experienced pollution problems similar to those of Eastern Europe. However, from the 1970s onwards, they began to adopt pollution amelioration policies and many Western European firms underwent a process of ecological modernization. In contrast, while environmental legislation was strengthened in Eastern Europe during the 1970s and 1980s, in some cases setting standards above those in the West, little attention was paid to actual implementation. Ecological modernization also failed to take hold. The capacity of Eastern European countries to deal with their pollution was limited, not least because their environmental infrastructure suffered chronic neglect under the old regimes. Furthermore, firms had to give priority to reaching production quotas, not meeting environmental standards. Factory managers were often closely associated with the local political elites and were often the major employer in a town or area, enabling them to command considerable power at the local level. Firms, particularly large firms, were often put in charge of their own environmental monitoring. Environmental fines were set at very low levels, often making it more 'rational' to pay fines than to install costly pollution prevention measures. In other words, under the old communist system there was a very close relationship between the economic and political elites and the system of public administration. This embeddedness resulted in only weak responses to the growing environmental problems that industrialization was causing in the region (Baker 2002; Pickvance 2004).

The environmental situation, however, has not remained static since 1989. Governments have undertaken environmental clean-up, particularly in heavily polluted industrial zones. The collapse of production in many of the large state-owned industrial enterprises has also led to improvements in ambient quality. There has been, for example, a particularly noticeable improvement in air quality in the Czech Republic (Carter and Turnock 2001). Similarly, the industrial restructuring that has occurred since 1989 has discouraged the high levels of resource use typical of the old system. As a result, the renewed industrial activity seen since the late 1990s has not resulted in a return to the low environmental standards of the past.

However, there is no simple correlation between the end of communist rule and improvements in environmental quality. The period since 1989 has witnessed a reduction in some forms of environmental pressure, only to find this accompanied by the emergence of new environmental problems. Examples include the growing problem of consumer waste and packaging as well as the problems associated with the rise of road transport and the use of private cars.

Improving environmental management

Attention has also been paid to enhancing the legal, administrative and institutional capacity – the environmental governance capacity – of the state to manage the environment. The largest single influence on that process has been the desire of countries in the region to become members of the EU. This is discussed below. This influence is acting in parallel with the development of new strategies for regional cooperation, especially between neighbouring states that share a common ecological feature or resource, such as a river. This includes cooperation among countries in the Danube river basin and those bordering the Black and Caspian seas, a process financially aided by the EU (Baker 2005a).

Promoting ecological modernization

In addition to these direct, state-level responses, there has been a parallel process of ecological modernization of the economy, which is shaping the promotion of sustainable development in transition societies. Much of this is taking place at the level of the firm and, in particular, within the production process. The privatization of state-owned companies has provided an important conduit for the transfer of ecological modernization practices to the industrial sector. Privatization, especially when it has involved purchase by foreign companies, has been followed by plant modernization, which has helped to improve air quality, reduce the energy intensity of production and improve waste management and resource recovery practices. The Belgian company Union Minière, for example, which bought a controlling stake in the Bulgarian Pirdop metallurgical plant, has installed new dust abatement technology and modernized waste storage and management in the plant (US Geological Survey 2000).

However, while ecological modernization can make an important contribution, the promotion of sustainable development involves a broader range and deeper set of social, economic and political changes. In addition, there is no guarantee that privatization will result in ecological improvements. In some countries, particularly in the Balkans, including Bulgaria, privatization has not been a force for ecological modernization. On the contrary, state assets have often been sold into the hands of the old *Nomenklatura*, which has a history of eschewing environmental regulations. Here privatization, instead of being a modernizing project, has enabled the old political elites to gain new, economic, power. It has also allowed the political 'embeddedness' of institutions of governance, industrial production and environmental regulation to endure in the post-communist period (Baker 2002).

Enhancing democratic participation

The transition process has also brought profound changes at the political and social levels. The formation of political parties, the holding of regular, free elections and the establishment of several new state environmental bodies and institutions all allow environmental concerns to be routed through the democratic system and its newly forming institutions and structures (Tickle and Welsh 1998). Democratization and political modernization have also helped to introduce greater transparency into the environmental policy-making process. Decentralization of public administration has given local authorities a new voice in environmental management, while at the same time exposing their poor resources and weak administrative capacity.

Yet, while the period since 1989 has seen the democratization of political life and some ecological modernization of industry, the social conditions necessary for the promotion of sustainable development have, in some respects, deteriorated. There is a growing gap between those sections of the community that are benefiting from the economic opportunities of transition and those that are being marginalized by it (Baker and Welsh 2000). More generally, the emergence of a class of *nouveaux riches* and of organized crime has the potential to threaten the environmental gains won by transition, since these constitute the least public-spirited segments of society. Such groups demonstrate a preference for private gain over the common good, and their continued strength retards the formation of strong environmental norms in the region (Baker 2005b). Low priority is given to environmental protection compared with economic development, especially in the face of EU membership. Furthermore, despite the enhancement of the role of environmental NGOs in public policy formation and implementation, there are nevertheless problems with the on-going weaknesses in civil society and the continuation of centralized administrative structures alongside closed and highly politicized bureaucratic cultures (Smith and Pickles 1998). These characteristics are not in keeping with the types of structures, institutions and processes needed to promote sustainable development. At the same time, marketization and the growth of consumerism have given rise to a new wave of environmental problems associated with consumer waste and the growth of private car ownership.

Sustainable development and EU membership

Eight Eastern European countries joined the EU in May 2004. The so-called 'first wave' member states are the Czech Republic, Hungary, Poland, Slovakia, Slovenia and the three Baltic states of Estonia, Latvia and Lithuania. Two other accession countries, Bulgaria and Romania, are expected to achieve membership of the EU in 2007.

In all these countries, membership of the EU became the determining factor shaping environmental policy in the post-1989 period. It is fair to say that the influence of the EU extends beyond these countries, to the entire Eastern European region. An example of this is the EU's role in shaping structures of regional environmental cooperation, as mentioned above. The EU also plays a key role in the Environment for Europe Process, a pan-European cooperation for environmental management. As a result, the prospects for the promotion of sustainable development in transition countries are closely linked with preparing for, and attaining membership of, the EU. This is not to deny that countries in the region also struggle to develop their own, indigenous responses. In addition, there is some evidence to suggest that Russia, despite the low priority given to the environment, is beginning to draw upon indigenous understandings of the relationship between society and the environment to construct a particularly Russian interpretation of sustainable development (Oldfield and Shaw 2002), which also has the potential to influence developments in the region. While it would be foolish, even arrogant, to assume that Russia, and indeed the region as a whole, had nothing to teach the West about conservation and environmental management, the EU is exercising, none the less, a predominant influence in the region. Because of its importance, attention is now turned to the exploration of the promotion of sustainable development in transition states within the context of preparation for, and membership of, the EU. Did preparation for membership of the EU have a positive or negative influence on the prospects for sustainable development in the region?

Strengthening environmental legislation

As part of their preparations for membership, countries had to adopt the *acquis communautaire* of the EU – that is, the entire body of EU legislation, treaties and case law. The adoption of the *acquis* strengthened the environmental legislation of the countries and it broadened the range of issues covered by legislation and policy (for example, waste management). There is also new pressure to achieve more effective policy implementation, often only weakly addressed under the old communist system. Several of the new member states have made explicit commitments to the promotion of sustainable development, which is enshrined in EU treaties and forms part of the environmental *acquis*. The Estonian parliament adopted the Act on Sustainable Development in 1995. The principles of sustainable development formed the basis of the 1995 Polish National Environmental Policy; Poland also has a Council for Sustainable Development. Similarly, Hungary established a Commission on Sustainable Development in 1993.

Nevertheless, there are major challenges ahead, involving, for example, the regulatory components needed to promote sustainable development. Adoption of the environmental *acquis* is but one step – legislation has also to be implemented and enforced. Environmental legislation is both expensive and technically complex to implement. The new member states have to update, extend or build installations and infrastructure, such as wastewater treatment plants, in order to be able to implement the Urban Wastewater Directive. Lack of administrative and financial resources also hampers effective implementation of legislation, and monitoring remains weak. The Commission has estimated that the new member states will need around €80 billion to €110 billion investment to conform to EU environmental legislation, or around €1,057 *per capita*. For example, Polish compliance with the Urban Wastewater Directive alone will cost almost €7 billion (CEC 2001e).

The impact of promoting sustainable development in transition societies can be explored by looking at the pressures that the transition process is placing on the sectoral level – for example, on the transport, agricultural and industrial sectors.

Box 8.1 EU pre-accession funds

The EU provided financial assistance to help countries prepare for membership. The three main pre-accession funds were the:

- *Instrument for Structural Policies for Pre-accession* (ISPA): is the principal means through which the EU provides environmental aid. It funds both environmental and transport infrastructure developments, targeting 'investment-intensive' environmental directives.
- *Special Accession Programme for Agriculture and Rural Development* (SAPARD): supports 'structural adjustment' in the agricultural sector and rural areas, and has an environmental component. Structural adjustment is taken to mean improvements in technical infrastructure, modernization of the farming sector and intensification of agricultural production.
- *PHARE*: launched in 1989, the PHARE programme provides finance for economic and political reform. Much of its budget is now devoted to institutional capacity building.

Sources: adapted from FoE Europe and CEE Bankwatch Network (2000); CEC (2002b); CEE Bankwatch Network and FoE Europe (2002).

The next section explores if, and to what extent, environmental considerations were integrated into pre-accession policies by looking at developments in five key sectors and in nature and biodiversity conservation policies.

Transition, EU membership and the challenge of environmental policy integration

The agricultural sector

Unlike in their Western European counterparts, agricultural practices in the new member states have largely worked in cooperation with nature and the landscape. Mixed farming and low-intensity agriculture created habitats for many species of wild plants and animals. The region has maintained a rich and diverse landscape, which ranges from coastal meadows and wet grassland in the Baltic region to the strip-land farming landscape of southern Poland and small-scale livestock rearing in the Carpathian mountains. As such, agricultural and rural areas represent one of the most significant contributions, in terms of natural capital, cultural heritage as well as social cohesion, which the new member states bring to an enlarged Europe (FoE and CEE Bankwatch Network 2000). Building upon this natural and cultural capital would provide a good stepping stone in the promotion of sustainable development.

This challenges the EU to reform its agricultural policy so that it can value nature as a key component of the wealth of rural Europe (WWF 2002a). However, the Common Agricultural Policy (CAP) of the EU, despite leading to serious deterioration of the rural environment of Western Europe, is being applied in the new member states. Despite several reforms that brought environmental considerations into the CAP, the logic of EU agricultural policy remains production-orientated, and a switch towards more ecologically responsible land and landscape management has yet to take place. As a result, the CAP still exercises negative pressure on Europe's environment, and is responsible for on-going problems of soil erosion, water pollution and biodiversity loss.

The low-intensity agricultural practices of Eastern Europe also offer an opportunity for the transfer of best practice from Eastern Europe to the West. This is very relevant, as Western Europe struggles to de-intensify agriculture in the face of the rising financial cost of the CAP, agri-chemical pollution, food safety scares, growing concern about animal welfare and husbandry practices and loss of biodiversity. There is the possibility of synergy between CAP reform, agricultural development under transition and the creation of a new model of rural sustainable development in an enlarged EU. However, the fate of rural areas following the last enlargement in 1995, when Austria, Finland and Sweden joined the EU, bodes ill for the future. For example, after member-ship, the small-scale farms in the forest areas and the archipelagos of Sweden, particularly northern Sweden, suffered from increased abandonment, with the loss of many species and habitats listed under the EU Birds and Habitats Directives (EEA 2003). The traditional production-orientated approach of the CAP that lies behind this abandonment is reflected in current pre-accession agricultural strategies, particularly the SAPARD programme. SAPARD supports agricultural intensification, despite the resulting pollution problems and habitat loss it can cause. It does not target improvements in quality of life and living conditions in the context of rural development. This means that the promotion of the social pillar of sustainable development through, for example, maintaining or enhancing social cohesion, does not receive attention. Given current practices, the opportunity to draw upon the practices of Eastern and Central European countries to help construct new models of sustainable agriculture in Europe will be lost.

The energy sector

The energy sector has an impact upon the environment at three levels: at the local level, by releasing particles and smog into the air, for example; at the regional level, by producing acid rain, for example; and at the global level, by inducing climate change, for example. These effects have major negative consequences for human health and biodiversity. Thus it is fair to say that the way in which the energy sector is developing in transition countries provides a litmus test for ascertaining whether or not the transition process is helping to promote sustainable development in Europe.

The energy strategy of the Commission for the new member states has several components (Box 8.2). Implementing this strategy will not be easy. The energy intensity of the economies in the new member states is more than three times higher than the EU average (CEC 2003). The sector still uses outdated technology and relies on poor-quality fuel. Nevertheless, these difficulties are also an opportunity. Several of the Commission's priorities could contribute to the promotion of a sustainable energy policy and, hence, to the promotion of sustainable development in an enlarged Europe.

Box 8.2 The Commission's energy strategy for new member states

- Constructing an efficient, effective and equitable energy policy.
- Creating the internal energy market and speeding up the liberalization process.
- Building up oil stocks.
- Restructuring or closing existing solid fuel (mainly coal) plants.
- Promoting energy efficiency and the use of renewable energy.
- Promoting co-generation (combined heat and power systems).
- Developing demand side measures.
- Ensuring nuclear safety.

Source: adapted from CEC (2001b).

There is strong synergy between the planned reduction in energy intensity by 1 per cent per year until 2010, the achievement of greater energy efficiency through the ecological modernization of the energy sector, particularly with respect to production, and an enlarged Europe meeting its Kyoto Protocol targets. The emphasis on renewable energy is also important, as is the extension of existing energy efficiency programmes to the new member states. Latvia has already shown leadership in this area, most if its electricity coming from hydro and co-generation sources. In Latvia, however, the development of numerous small hydropower stations threatens to disturb river basin management, with a knock-on impact on protected species and important habitats. This indicates the importance of ensuring that the shift to renewable energy is also sensitive to broader environmental considerations.

Similarly, restructuring coal-fired power stations could reduce greenhouse gas emissions. Old and inefficient coal-fired plants contribute almost half of all CO_2 emissions from the region. There is real potential for reducing greenhouse gas emissions through the ecological modernization of power plants, transmission and grid systems, and through the application of new coal technology. This could help improve urban air quality in Bulgaria, especially in the capital, Sofia, which burns brown coal (lignite) with a high sulphur content. It is also important for the switch from reliance upon unsafe nuclear technology. Ensuring nuclear safety is an essential prerequisite for the preservation of our collective future (Box 8.3).

Box 8.3 Nuclear safety

A 1998 EU-sponsored study of nuclear safety in Eastern and Central Europe found that six reactors were operating at high levels of risk. Reactors in Bulgaria, Lithuania and Slovakia were classified as 'non-upgradable at reasonable cost'. These included the Kozloduy reactor in Bulgaria, the Ignalina reactor in Lithuania and the Jaslovské-Bohunice reactor in Slovakia. There was deep concern about the Ignalina reactor, because it is of the same design, and uses the same technology, as the Chernobyl reactor in Ukraine.

In 2000 Bulgaria agreed to close some of the Kozloduy units, in exchange for a substantial loan from the European Commission to modernize the remainder of the nuclear plant. However, progress in meeting this commitment has been slow. There is little evidence that Bulgaria is either strengthening the capacity of its Nuclear Regulatory Authority or developing alternative energy strategies, needed as part of the plans to close down units of Kozloduy.

Lithuania's Ignalina reactor is due to be decommissioned by 2009, and closures are expected at Slovakia's Jaslovské-Bohunice reactor. A leak in April 2003 from Hungary's Paks nuclear power plant has added this facility to the list of controversial sites in the region.

Sources: adapted from CEC (1998d, 2002c); REC, *The Bulletin*, 12(1) May 2003.

However, the benefits of improvement may be outweighed by the fact that transport emissions are growing steadily, and road transport emissions are expected to double in the next two decades. These trends would suggest that the potential for 'decarbonization' of the economies of the region is weak. There are also substantial structural and operational obstacles to the integration of environmental considerations into energy policy. However, the chief obstacle remains ideological: the development of a sustainable energy policy is not a priority for many new member states, which view security of supply, or, more particularly, security of electricity supply, as more important. Historical factors may account for this, especially the electricity shortages experienced during the period of communist rule (Eichhammer 2001). Furthermore, the Commission is determined that the integration of environmental considerations into energy policy should also take account of the other priority goals of energy policy, such as competitiveness and security of supply (CEC 1998a).

Box 8.4 Sustainable energy policy in an enlarged Europe: the challenges

- Substantially increased energy demand is expected until 2020: electricity +50 per cent; residential +41 per cent; transport +30 per cent.
- CO_2 emissions are expected to grow: 1990–2010 +7 per cent; 1995–2020 +15 per cent.
- Import dependence will grow, especially with respect to oil.
- Building up oil stocks will impose a heavy financial burden, which has a high opportunity cost.

Source: adapted from Eichhammer (2001).

The industrial sector

As in the energy sector, there is also strong synergy between economic restructuring in transition societies and ecological modernization, particularly of production processes. Cleaner technology and improved environmental management are hallmarks of ecologically modernized firms, leading to reductions in resource use and to resource recovery in the production process. Ecological modernization is a key element in the EU's approach to the industrial sector of the new member states. Reducing the cost of production is seen as necessary for industrial competitiveness, especially for fledgling export-orientated firms (CEC 2001c, 2002d). There is also a belief that the reduction of energy and resource use in production can be of particular help to the small and medium-sized enterprises that are mushrooming in the region following the collapse of the old state-run command economy.

The pollution problems in many parts of Central and Eastern Europe were, in large measure, caused by the industrial sector, with its old and decaying installations and wasteful use of resources, especially energy use, in production. The collapse in production since the end of the old regime has helped in no small way to reduce pollution, especially air pollution, in industrial regions. However, the reduction in pollution is also due to investment in best available technologies (BAT), which has increased eco-efficiency in the sector (EEA 2003). The adoption of EU regulations and technical standards has also helped raise environmental performance across the industrial sector.

However, fewer improvements have been made in some of the more polluting manufacturing sectors, especially mining and chemicals. These sectors are experiencing high levels of growth. They are also the sectors where the technical improvement measures with lowest costs have already been implemented (EEA 2003), making it more difficult to achieve further environmental gains. There is also need to address the rise of industry-driven transport demand. Soil contamination from local industrial plants, caused by industrial accidents and improper industrial waste disposal, also remains a problem. Often these industrial plants are no longer in operation, making it difficult to apply the 'polluter pays' principle, or to deal with the legal problems of environmental liability. Furthermore, there is still need to put in place appropriate institutional and regulatory frameworks and improve the enforcement of environmental standards in the industrial sector.

The tourism industry has also seen rapid growth since 1989. This is a strong contributor to transport growth. In addition, tourism increases the demand for water, leads to the generation of local waste and to local land fragmentation and disturbs natural habitats. While tourism is increasing, policy measures to promote more sustainable tourism are progressing very slowly.

Resource and technical difficulties aside, there is a more fundamental issue: a stark choice faces this sector. The EU's industrial strategy for an enlarged Europe, known as the Lisbon Strategy, set the goal of making the EU, by 2010, 'the most competitive and dynamic knowledge-based economy in the world, capable of sustainable growth with more and better jobs and greater social cohesion' (CEC 2002d: 7). The European Environmental Agency has warned that the Lisbon Strategy is placing harsh demands on Europe's land resources. The agency's Executive Director, Professor Jacqueline McGlade, has argued that:

> These demands on the natural capital have spilled out well beyond Europe's boundaries. So much so that we must now face up to the realisation that to move forward on a trajectory designed to meet the Lisbon agenda, Europe will have no option other than to exploit the rest of the planet or fundamentally alter the way in which it does business by becoming dramatically more efficient in its use of land and other natural resources.

> (McGlade, in EEA 2004)

The challenge for the EU is to ensure that its commitment to sustainable development is strong enough for it to reduce its ecological footprint.

The forest and fisheries sectors

In many countries in Eastern and Central Europe a relatively high proportion of forest territory has Protected Area status. The period of communist rule proved particularly good for the maintenance of rich forest landscapes across the region, resulting in high levels of biodiversity. In both Hungary and Slovakia, for example, approximately 20 per cent of the forest is within Protection Areas. In some countries, legislation is stronger than that provided by the EU *acquis*. This could have been viewed as a positive asset in negotiations between the EU and the candidate countries, especially given that the EU is developing competence in this area but lacks experience and expertise. Unfortunately, such was not the case and there is a real danger that membership of the EU will reduce environmental standards in the forest sector in the region.

While it is evolving 'creeping competence' in relation to the forestry sector, the EU has no adequate policies to promote sustainable forestry practices. Furthermore, there is a lack of coordination between policy in this sector and those in closely related sectors, including the EU Biodiversity Strategy. The Commission also lacks a clear strategy on how to handle forests within an enlarged EU. When account is also taken of the policies of privatization and of land restitution, it is clear that the forests of Eastern Europe are becoming very vulnerable to non-sustainable forms of commercial exploitation. Lithuania, for example, has agreed new logging concessions as commercial forestry takes hold in the country.

The EU's Common Fisheries Policy (CFP) was designed to resolve conflicts among member states over territorial fishing rights and to prevent overfishing. Eastern enlargement has extended the EU's coastline and seaboard territory, adding significant additional waters to the jurisdiction of the CFP, in the Mediterranean, Baltic, Adriatic and Black seas. Fish stocks are heavily depleted in these seas, and the Black and Baltic seas suffer from significant levels of pollution (Carter and Turnock 2001).

The outlook for an enlarged Europe remains pessimistic, as most CFP reforms have been driven by the commercial crisis of fishing caused by over-exploitation and not by environmental considerations. Much needs to be done still to support the conservation and sustainable use of commercial stocks and marine eco-systems, especially as efforts to date have not halted the decline in fish stocks. There is also little coordination between the CFP and the various seas conventions that apply in the region. These include the 1978 Convention for the Protection of the Marine Environment and the Coastal Region of the Mediterranean (the Barcelona Convention), which was revised in Barcelona in 1995 and is not yet in force; the 1992 Convention on the Protection of the Marine Environment of the Baltic Sea Area (the Helsinki Convention); and the 1994 Convention on the Protection of the Black Sea against Pollution (the Bucharest Convention). However, while the region is subject to several international legal marine conventions, its marine habitats and species are underrepresented in the annexes of the Habitats Directives (WWF 2000), which means that they may not be adequately protected under EU law.

The transport sector

EU transport policy has long been a key source of environmental stress, particularly with the large-scale infrastructure projects, especially road building, introduced as part of the European single market programme. The integration of environmental considerations into these projects has been slow, 'sustainable transport' remains poorly conceptualized in policy terms and long-term targets have not been developed (CEC 1999c). The problem, however, lies not just with the EU. Policies to decouple growth in transport emissions from growth in GDP have yet to be developed. Transport presents a *persistent* environmental problem, one that is not amenable to readily available solutions. This means that the development of a sustainable transport system requires deep cultural changes. They include changes in work patterns, lifestyle, leisure activities, and in the principles that guide land use and spatial planning. Bringing about such changes is a slow process, people have to be given incentives to change, but there also has to be a credible threat of regulatory intervention. In other words, the development of a sustainable transport policy has to conform to ideas of good governance and democratic participatory practice, while at the same time taking account of the need for strong government intervention in this area. Most member states are not committed to this practice.

The development of sustainable transport systems within the transition states could start by utilizing their existing advantages: most new EU member states still have a rail share well above the EU average, lower transport energy use, lower pollution emissions *per capita* and less fragmentation of their land from transport infrastructure, particularly roads. However, transition is bringing modernization and economic growth, and governments are under pressure to meet societal needs for improved standards of living. All these pressures bring increased mobility demands, especially for private transport. It has also been argued that the transfer of the Western culture of 'freedom, independence and privacy' is another element contributing to the rise of private car transport, and the associated road-building programmes, in Eastern Europe. This is blocking measures to stimulate other models of sustainable transport (Grin *et al.* 2003).

Like their Western European counterparts, Eastern European countries lack a comprehensive model that can help them decouple the growth in transport demand from their economic and social development. However, experience in the West has shown that, while a comprehensive model is lacking, certain facts are known: road-building programmes increase road congestion, enhance private car dependence and cause high levels of pollution, both locally and in terms of their contribution to climate change. Such infrastructural investments also have high opportunity costs, in that they deflect resources away from the development of more environmentally friendly community transport systems.

The transition process, instead of maintaining existing transport advantages, has instead led to worrying trends. Transport volumes are increasing significantly, particularly road transport, in part owing to the rise of east–west trade. More significantly, membership of the EU has meant that the trans-European transport network (TEN-T) has been expanded eastwards. This major pillar of the EU's Common Transport Policy mostly involves large-scale motorway building projects. The ISPA pre-accession fund finances this expansion. However, there has been no strategic environmental impact assessment of the TEN-T, or of its extension to the new member states or the accession countries.

Box 8.5 Trends in transport in new member states and in accession countries

Environmental performance of the transport sector

- Energy consumption by transport is increasing rapidly, mainly due to the growth of road transport.
- Transport CO_2 emissions dropped in the early 1990s, but are now growing with traffic volumes.
- Land take by transport infrastructure is increasing.
- Land fragmentation, while less than in the EU, is increasing with infrastructure development.
- Infrastructure developments are adding to the pressures on designated nature areas.
- The numbers of 'end of life' vehicles and used tyres are expected to grow significantly.

Management of transport demands/modal split

- Freight transport is shifting to road.
- Passenger transport is shifting to road and air.
- Motorway lengths have doubled in ten years.

---

continued

- Investment patterns indicate that road building is given priority.
- Fuel and transport prices are not environmentally sensitive.

Environmental management

- Integrated transport and environment strategies are lacking.
- Institutional cooperation is seldom formalized.
- Monitoring of environmental policy integration in transport policy is lacking.

Source: adapted from EEA (2002b).

At times, it is difficult to distinguish the impact that the preparation for membership of the EU is having on the environment from the more general impact of transition, which is bringing modernization, a growth in consumerism, market liberalization and economic restructuring. However, in relation to transport policy, there is a clear link between the transport infrastructure development policy of the EU and enhanced pressure on the environment in Eastern Europe, particularly upon natural habitats (Box 8.6).

Box 8.6 Impact of ISPA-funded transport development on the natural environment of Eastern Europe

The ISPA pre-accession fund of the EU funds both environmental and transport infrastructure and development. This dual mandate is controversial, leading to criticisms that the ISPA is a weak tool for environmental management. Its environmental safeguards are weaker than those of the EU Cohesion Fund, on which the ISPA was modelled, despite the fact that the Cohesion Fund had been subject to severe criticism on environmental grounds. Research by both Friends of the Earth and CEE Bankwatch Network suggests that the ISPA prioritizes road building over environmental protection; it has been accused of developing a 'car dependent' society in the region.

EU funding has been specifically criticized for its support of the VIA Baltica TEN-T motorway, which cuts through the Biebrza National Park in Poland, and the Struma motorway, which would pass through the length of the Kresna gorge in Bulgaria, a site marked for inclusion in the Natura 2000 network. Similarly, a section of the EU-funded D8 motorway, part of a European transport corridor, passes through a nature park, a site also being prepared for inclusion under Natura 2000.

BirdLife International has also examined the impact of ISPA funding on the region. It found that eighty-five important bird areas (IBAs), or 21 per cent of all IBAs investigated, would be negatively affected by the eastward expansion of the TEN-T, particularly its road-building programme.

The project of Slovakia and the Czech Republic to develop the Danube–Oder–Elbe canal, with its related Oder 2006 dam building project, is also problematic.The canal threatens some twenty-six potential Natura 2000 sites on the Polish side of the river and two sites in the Czech Republic. The ISPA funds the Odra 2006 plan. All these developments are subject to on-going controversy and have been the focus of attention of environmental NGOs.

Sources: adapted from BirdLife International (2004); Friends of the Earth Europe and CEE Bankwatch Network (2000); WWF (2003).

Nature protection and biodiversity conservation

Because of fewer development pressures, lower-intensity agriculture and their system of forest protection, the new member states host species and habitat types that have nearly vanished from Western Europe, including mammals such as the European bison. These countries have the last great wilderness areas on the European continent: the Carpathian mountains, stretching across seven countries in the region and one of the last homes of large European carnivores, including bears, wolves and lynx; the Danube delta, globally significant as a wetland area; the river Vistula in Poland, one of the large European rivers with major natural features; and the Baltic coast, one of the most important corridors for migrant birds in Europe (WWF 2003). This rich natural and cultural heritage constitutes significant 'comparative advantages', since 'they form the potential foundation of development strategies based upon the principles of sustainable development' (WWF 2000: 3). For this to be realized, however, the comparative advantages of the region have to be maintained. Natura 2000 is the chief instrument that the EU has at its disposal for this task.

Natura 2000

Two key pieces of legislation frame EU nature protection policy and policy aimed at the conservation of biodiversity. The first is the 1979 Directive on the Conservation of Wild Birds (the Birds Directive) and the second is the 1992 Directive on the Conservation of Natural Habitats and of Wild Fauna and Flora (the Habitat Directive). The Habitat Directive is potentially the most important of the EU instruments. It led to the establishment of the Natura 2000 programme, aimed at the creation of a comprehensive, linked network of protected European habitats (Baker 2003).

New member states have to implement both the Birds and the Habitat Directives and implement the related Natura 2000 programme. However, the eastern extension of the Natura 2000 programme has proved difficult. To begin with, nature conservation is a policy area plagued by disputes over whether the EU should have competence here or whether it is an issue best left to the member-state level (Baker 2003). Consequently, the implementation of the Natura 2000 programme is already subject to considerable delay in existing member states. The eastward expansion of Natura 2000 has exacerbated this problem.

The Commission has also tended to treat the eastern expansion of the Natura 2000 programme as a mere 'add-on' to the established processes and frameworks in place for nature protection in Western Europe. This has resulted in failure to take account of the very different assemblage of species and habitats, especially the many endemic plants and animals, in the regions. One of the results is that species native to Eastern Europe are underrepresented in the annexes of the Habitat Directive, which means that they may not be adequately protected under EU law. A similar problem with respect to marine species was noted in our earlier discussion of fisheries (pp. 201–2). In addition, there is a presumption that little attention needs to be paid to what the EU itself might learn in terms of nature conservation measures from the low-intensity agricultural practices and the forest management strategies of the new member states.

However, responses to date within the new member states can also be criticized. There is a common tendency for countries, when identifying Natura 2000 sites, to limit them to areas that already have protected status. This means that site identification often makes little reference to wider ecological regions or to the preservation of biodiversity in Europe as a whole (WWF 2002b). Furthermore, there is insufficient independent control over the site selection process.

In addition, all countries have acknowledged that they lack the administrative and institutional capacity to implement the Natura 2000 programme – that is, to manage, to administer and to monitor sites. Slovakia seems to be the only country where the financial and human resources available for nature conservation have increased since 2000. This follows the establishment of the State Nature Conservancy. While there is some funding available from the EU, through the EU LIFE programme, for example, the main financial instruments for pre-accession, the ISPA and SAPARD do not fund Natura 2000. There are also attempts to enhance administrative capacity through 'twinning arrangements' – that is, where experts from government agencies and bodies in older member states work with their eastern counterparts to teach new skills, procedures and management prac-tices. There is a twinning project, for example, between the Polish Ministry of the Environment and those charged with the administration of French regional parks.

Another major difficulty with the pre-accession funds is the mismatch between the timing of investment funds for traditional development activities, such as road-building programmes, and those designed to help nature conservation. Many of the decisions concerning the use of the three pre-accession funds were made before the required nature conservation measures (the Birds and Habitat Directives, Natura 2000) were put into place. In practical terms, this meant that transport network developments (motorways, waterways, etc.) funded through the ISPA and agricultural developments (intensification, modernization) funded through the SAPARD proceeded *in advance* of the steps necessary to safeguard biodiversity. As a result, when the nature conservation systems are finally in place, they will help conserve only the biodiversity that has managed to survive large-scale investment programmes aimed at promoting Western forms of development (WWF 2000). Older member states made the same mistake in the past, especially those in peripheral regions, such as Ireland, which received large amounts of EU structural funding and which are now able to implement nature conservation strategies to preserve only what is left over after modernization and development have taken place. This despite the realization that the promotion of sustainable development requires the application of the principle of environmental policy integration – that is, the integration of environmental considerations into all stages of the policy process.

Many of the problems that have been discussed in relation to nature conservation strategies in transition economies are of a technical nature: they relate to the management and strategic use of administrative, financial and technical resources. It is clear that inter-ministerial coordination is poor in many of the new member states. There is a lack of mechanisms for facilitating joint decision-making processes, and institutional arrangements for policy coordination are weak. These are basic conditions for effective environmental policy integration. The controversies also point to the fact that land-use planning is poorly reflected in Natura 2000. In short, there has not been adequate supervision of the environmental impact of EU pre-accession policies. As such, new member states are rather like their Western European counterparts – both groups find it very difficult to adhere to the principle of environmental policy integration.

However, the problems facing the transition societies of Eastern Europe are not confined to mere technical, administrative matters. A structural problem exists, in that the environment and social cohesion of Eastern Europe are under increased pressure from the development process itself that has been under way since the collapse of the old communist regime. It would appear that neither the new member states nor the EU have successfully responded to the challenge of guiding transition in ways that use and profit from the region's natural heritage. Rather, development is undermining or destroying that heritage. In short, unsustainable

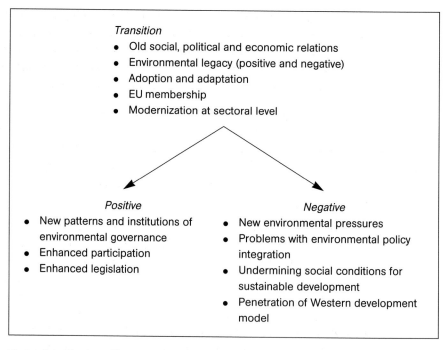

Figure 8.1 *The transition process, while enhancing environmental management capacity, fails to grasp sustainable development opportunities*

patterns of development are taking hold in Eastern Europe. The challenge of environmental policy integration lies in counterbalancing these patterns with new principles of development that combine social cohesion, ecological integrity and economic progress into a new sustainable synthesis.

Conclusion

This chapter explored the prospects for, and barriers to, the promotion of sustainable development in the transition countries of Eastern Europe. The environmental dimensions of transition have to be viewed within the context of the complex interface that is evolving between the social, political, cultural and administrative legacies of the old regimes and the new systems of environmental management that are being introduced across the region. On the one hand, there is a great deal of continuity with the past, especially the low priority given to environmental protection over economic development. At the same time, marketization and the growth of consumerism have given rise to a new wave of environmental problems associated with consumer waste and the growth of private car ownership. The role played by environmental NGOs in public policy formation and implementation, although expanded since 1989, remains limited by the on-going weaknesses in civil society and the continuation of centralized administrative structures alongside closed and highly politicized bureaucratic cultures. These characteristics are not in keeping with the types of structures, institutions and processes needed to promote sustainable development.

On the other hand, transition has seen the emergence of new features of environmental governance, in part driven by external influences. Countries in the region, especially those that have joined the EU, have become anchored into the system of international environmental governance. This has helped spread a commitment to the principles of sustainable development, at least at the level of declaratory politics.

The chapter also explored one of the principles of sustainable development: environmental sectoral policy integration. It examined the application of this principle to the accession process – that is, the preparations made by several transition states for membership of the EU. At its simplest, environmental policy integration requires an end to contradictory policies. Like their counterparts in existing member states, governments in the new member states give low priority to environmental protection, nature conservation and, more generally, the promotion of sustainable development, when faced with the challenges of modernization and development. The threat of competition and the drive to raise production levels in advance of EU membership increased pressure towards unsustainable development in accession countries. Economic growth is already changing consumer habits and generating more waste. Transport is expected to increase by as much as seven times its current volume, and most of it will be road traffic, one of the most polluting forms of transport. Public transport networks are declining, and an explosion of waste is filling landfills in countries that had good systems of resource recovery and recycling. The social conditions necessary for the promotion of sustainable development are deteriorating. These trends, combined with expanding tourism, more intensive farming and forestry, and the expected increases in energy consumption, have real potential to destroy the invaluable natural environment that exists in the region. The tragedy of this development is that it is funded and aided by the EU, despite the fact that the EU has both a declaratory and a legal commitment to the promotion of sustainable development.

The negotiation process and preparation for enlargement were seen primarily in terms of candidate countries 'catching up' with the more advanced practices of their Western European neighbours. It became a chance to vindicate the Western form of development. There is a danger that the end of communism will result in the triumph of consumerism.

However, a different way of viewing the process of enlargement is to see the eastward expansion in terms of opportunity. Enlargement can bring enormous additional natural capital and biodiversity for the enlarged EU to cherish, enjoy and safeguard for future generations. It provides the social and ecological conditions for Europe to embark on a more sustainable path. From a global perspective, an enlarged EU could have a stronger, positive influence on efforts to address global environmental problems, such as climate change and biodiversity loss. 'Enlargement offers the European Union an opportunity to put its paper commitments to sustainable development into actual practice' (WWF 2003: 10). However, this chapter has shown that, far from taking advantage of the opportunity that enlargement brings, an enlarged Europe will be an environmentally poorer region.

Summary points

- Since the collapse of the communist regimes in Eastern and Central Europe in 1989, countries in the region have undergone complex processes of transition.
- The legacies of the old regime shape the prospects for, and barriers to, the promotion of sustainable development in Eastern and Central Europe. On the one hand, there is much in common with the past: low priority assigned to environmental protection; limited involvement of civil society in public policy; continuation of centralized administrative structures alongside closed and highly politicized bureaucratic cultures.
- Transition is bringing new features of environmental governance and anchoring the region into the system of international environmental governance.
- The social and ecological conditions for promoting sustainable development have deteriorated, especially with respect to social cohesion and ecological diversity.
- The EU plays a decisive environmental role in the region.
- Eastern enlargement has the potential to enrich the EU, ecologically and socially.
- Preparations for enlargement undermined the very significant ecological and social contributions that this region could make to the promotion of a sustainable Europe.

Further reading

Theoretical analysis of transition

Baker, S. and Welsh, I. (2000) 'Differentiating Western influence on transition societies in Eastern Europe: a preliminary exploration', *Journal of European Area Studies*, 8: 79–103.

Smith, A. and Pickles, J. (1998) *Theorizing Transition: The Political Economy of Post-Communist Transformations*, London: Routledge.

General overviews of the politics and policy of the environment in Eastern and Central Europe

Baker, S. (2005) 'The environmental dimensions of transition in Central and South East Europe', in I. Bell (ed.) *Central and South-Eastern Europe 2005*, London: Europa Publications.

Carter, F.W. and Turnock, D. (eds) (2002) *Environmental Problems of East Central Europe*, 2nd edn, London: Routledge.

Social aspects of the environment in Eastern and Central Europe

Tickle, A. and Welsh, I. (1998) 'Environmental politics, civil society and post-communism', in A. Tickle and I. Welsh (eds) *Environment and Society in Eastern Europe*, Harlow: Longman.

EU's policy towards the accession states

CEC (1998) *Accession Strategies for Environment: Meeting the Challenge of Enlargement with the Candidate Countries in Central and Eastern Europe*, Brussels: Commission of the European Communities.

European Community (1997) *Agenda 21: The First Five Years: European Community Progress on the Implementation of Agenda 21, 1992–1998*, Brussels: European Community.

Regional Environmental Centre (2003) *Environmental Financing in Central and Eastern Europe, 1996–2001*, Budapest: Regional Environmental Centre.

NGO reviews of EU policy towards accession states

CEE Bankwatch Network and FoE Europe (2002) *Billions of Sustainability: Lessons Learned from the Use of Pre-accession Funds*, London: CEE Bankwatch Network and FoE Europe.

Fisher, I. and Waliczky, Z. (2001) *An Assessment of the Potential Impact to the TINA Network on Important Bird Areas (IBAs) in the Accession Countries*, Sandy: BirdLife International/RSPB.

Friends of the Earth Europe and CEE Bankwatch Network (2000) *Billions for Sustainability? The Use of EU Pre-accession Funds and their Environmental and Social Implications*, London: Friends of the Earth Europe.

WWF (2002) *The 'New' European Union: WWF Agenda for Accession: An Update*, Brussels: WWF Europe.

WWF (2003) *Progress on Preparation for Natura 2000 in Future EU Member States*, Brussels: WWF Europe.

Web site

http://europa.eu.int, official EU portal.

9 Conclusion

The promotion of sustainable development. What has been achieved?

Constructing a new development paradigm

The exploration of the prospects for, and barriers to, the promotion of sustainable development began with an examination of environmentalism as a critique of the conventional model of development. Exploring the critique served to differentiate the model of sustainable development from the conventional approach. This helped to clarify what the model of sustainable development is designed to promote. The exploration outlined how the rise of environmentalism led to a fundamental questioning of the basic tenets of the Western development model, in particular as pursued in the post-war period of modernization.

Environmentalism criticizes the conventional model of development because it threatens the bases upon which future development depends. From an ecological point of view, this threat has become evident in biodiversity loss, climate change, deforestation and desertification and water shortages. When judged from a social point of view, deteriorating environmental quality causes social impacts that can weaken social and political stability. This decreases social cohesion and under-mines the assumption of a continuous, more or less harmonious development for society. Based on these critiques, it is no longer possible to see development in isolation from its ecological and social consequences. Environmentalism also leads to rejection of the idea of equating human progress with the domination of nature. The conventional model is also criticized for relying upon markets to distribute goods and services. Market access depends upon the ability to pay, while reaping the profits of production and service provision depends upon the on-going ability to commodify more and more areas of life (from food supply to leisure activities, from seeds to biological resources) by bringing them under the remit of the market system. Environmentalism rejects the equation of devel-opment with growth and discards the idea that consumption is the most important contributor to human welfare. More significantly, by showing that the model of development pursued by the Western industrial societies cannot be carried into the future, either in its present forms or at its present pace, environmentalism makes it imperative for society to construct a new development model.

The term 'sustainable development' forms the core organizing theme that integrates environmental, economic and social considerations into a new development model. The model is built upon normative principles that promote equitable access to the planet's limited resources in order to promote human needs, whether they are physical, cultural, spiritual or social. Equity extends across space – for example, between different geographical locations – as well as across time – for example, between generations – and operates across gender. In order to promote sustainable development, a halt has to be brought to the practice, typical within the conventional model of development, which allows the present generation to adopt a policy of *temporal displacement* – that is, to pass the risks and problems of modernity down to future generations. The *spatial displacement* of the negative environmental consequences of traditional development models has also to be stopped. Spatial displacement is a process whereby a more powerful state or actor imposes environmental harm upon another, less politically or economically powerful, state or actor (Blowers 1997). This can include the more powerful actor displacing industrial pollution or depleting the environmental assets, such as biodiversity, of another region or country for its own benefit. The stronger the form of sustainable development, the more weight is given to the additional commitment to sharing access between species – that is, between human and other life forms.

These normative principles drive a model of development that protects the planetary resources, whether they are physical, in the form, for example, of oil or gas, or systemic, in the form of the climate system, while it also promotes their use. It accepts a hierarchical interdependence between economy, society and nature: society is possible without a market economy, but neither society nor the market economy is possible without the natural environment.

While there are many, and often competing, versions of the model of sustainable development, they share the common belief that there are ultimate, biophysical limits to growth. Given these limits, to create the conditions necessary for ecologically legitimate development, particularly in the Third World, industrial societies have to reduce the resource intensity of production (sustainable production) and undertake new patterns of consumption, that not only reduce the levels of consumption but change what is consumed and by whom (sustainable consumption). The sustainable development model thus challenges conceptions of development that prioritize individual self-advancement. Rather, it holds that the promotion of the common good takes precedence over the encroachment on the commons by the few.

The sustainable development model has the following characteristics:

- Recognition of the value of the planet's biophysical and resource system.
- Imposition of limits on growth.
- Prioritization of the common good.
- Understanding development in terms of quality of life.
- Promoting socially and ecologically legitimate development, especially in the Third World.
- Reduction of consumption in the industrialized world.
- Acceptance of shared responsibility across multi-levels of governance.
- Participation in open-ended dialogue to identify and agree priorities.
- Respect for diversity as development trajectories are implemented across different social, cultural and ecological contexts.

Promoting strategic and political engagement

Understanding how the model of sustainable development works in practice involves studying the strategic and political contexts within which action to promote sustainable development takes place. Awareness of the *outer* limits of the earth's environment has gone hand in hand with a new awareness of the ways in which the *internal* organization of society, whether at the local or the international level, shapes the prospects for a sustainable future. Attention has thus to be given to the interlinked spheres of authority and influence that shape the way society is constructed and policies are made. These operate from the international down to the local level, from the transnational corporation to the individual level, and from the application of technology to the pursuit of a more spiritual engagement with nature.

Beginning with the international level, the involvement of the UN in the promotion of sustainable development is subject to two conflicting interpretations. On the one hand, the UN, particularly through holding environmental Summits, can be seen as having played a key, positive role in shaping an understanding of, and engagement with, the promotion of sustainable development. The Summits have helped the concept of sustainable development to permeate the official discourse not just of states but also of civil society and the economic sphere. The Rio Earth Summit was particularly important and the Rio Declaration has provided an authoritative set of normative and governance principles to guide development. Since Rio there has been progress in environmental institution building and in the development of new governance patterns, including at the national level. Today nearly all countries have government ministries and/or agencies in charge of the environment and several have established participatory

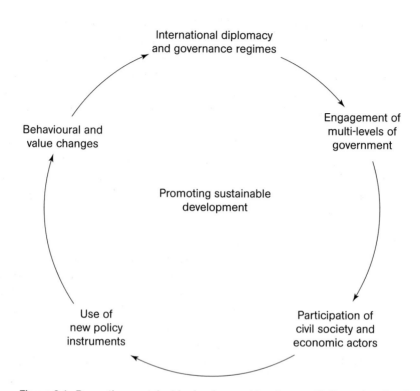

International diplomacy
and governance regimes

Behavioural and
value changes

Engagement of
multi-levels of
government

Promoting sustainable
development

Use of
new policy
instruments

Participation of
civil society and
economic actors

Figure 9.1 *Promoting sustainable development involves multi-dimensional actions*

sustainable development fora. This has led to the development of both hard and soft laws governing important ecological conditions for the promotion of sustainable development, across a range of issues from climate change to the transport of hazardous waste. There has been a proliferation of organizations from within civil society, and major social groups, including local authorities, business and industry, women and indigenous communities are now involved in the task. Most especially, at the local level, there has been an explosion of community activity under the banner of LA21.

There has been a shift in emphasis within the UN from the elaboration of principles to promote reconciliation between the economy and the environment to more practical considerations of implementing strategies. The WSSD epitomized this shift. In addition, new policy issues have been added to the sustainable development agenda. In particular, the growth of economic globalization has led to the increasing importance of trade in the promotion of global sustainable development, while simultaneously enhancing the role of international organizations such as the WTO.

Judged from this perspective, the UN can be seen as having made a positive contribution to the shaping of our collective future. It has structured the legal, institutional, political, economic and social engagements with sustainable development from the international down to the local levels. Nevertheless, it is fair to say that more progress has been made in environmental institution building than in actually protecting the environment or in implementing effective policies. There are also shortcomings in the institutions and the resources available for global environmental governance. The political will among UN member states to translate declaratory commitment into practical policies is also lacking. This has meant that the UNCED process is marred by the failure of its members to commit the necessary funds and by member states entering into negotiations driven by narrow, short-term and national interests.

The more positive view of the role of the UN runs alongside an opposite view, one that criticizes the organization as a management agent, helping to promote a system of global environmental governance that is preoccupied with means and not ends. Such preoccupation displaces a more fundamental critique of the flaws of conventional development policies, the structure of international politics and of Western-orientated environmental management practices. This approach questions whether the institutions of environmental governance that fall under the remit of UNCED can address the underlying causes of unsustainable forms of development.

Viewed in this more fundamental way, the challenge to promote sustainable development is not just about finding more effective and efficient institutions of environmental governance. It is also about genuine commitment to a common interest, developing new ecologically and socially based values and focusing on human rather than state security. It is ultimately about the distribution of power, between the global and the local, between the privileged and the marginalized, and about the priority given to the economic, the social and the environmental, at present and in the future. For many, the system of environmental governance promoted by the UN is incapable of addressing these more fundamental requirements. Viewed as such, the UN governance practices are seen as having opened up structures of governance without effecting changes in the processes of governance itself.

At the root of these conflicting interpretations lies deep conflict over whether sustainable development is a tool for the construction of radically different environmental futures or whether it should be rejected out of hand, as it represents little more than an anthropocentric management tool, useful to help capitalism to find a way out of its environmental crisis. For radical environmentalism, ensuring a sustainable future mandates the empowerment of the local and those most directly affected by environmental degradation as a way to hold public and private power and authority accountable. This requires new patterns of politics, including at the global level. For the less radical, market-led solutions can operate alongside technological innovations to promote a new era of development.

Ecological modernization: promoting weak sustainable development

The study of the engagement with sustainable development in different social, political and economic contexts has been particularly helpful in exposing the limitations of current practices. The commitment by the EU, for example, has allowed it to move beyond a policy approach that was dominated by the imposition of an ever tighter regulatory framework governing economic activity, especially production, to a new more constructive approach where environmental protection can be seen as a positive goal of economic activity. The EU's engagement has resulted in a shift in the understanding of the nature of the environmental problematic. It is no longer seen merely as an issue of pollution control and, hence, environmental regulation. Attention has shifted from earlier concern with resource management and ensuring that economic development resulted in an 'improvement in the quality of life' (decoupling) to an engagement with developing new forms of environmental governance (participatory, consensus driven). Despite these advances, new understandings of public, environmental citizenship that could result in dematerialization, especially with respect to consumption, are slower to develop. In addition, when the focus is moved beyond the declaratory commitments and treaty obligations of the EU to promote sustainable development, a less positive picture emerges. Both in relation to its recent eastern enlargement, and, more generally, in terms of its sectoral policies, the EU has promoted a weaker form of sustainable development, premised upon belief in the advantages of ecological modernization.

A model of development based on resource efficiency, pollution control and waste reduction cannot be generalized to the planet as a whole. Despite the advances of ecological modernization, growth still needs to be limited to come within the biophysical carrying capacity of the planet. Furthermore, consideration of equity principles lies outside the scope of the ecological modernization agenda. Corporate 'greening' is particularly suspect when considered from a Third World perspective, because it does not address issues of social justice and the equitable access to and distribution of resources (Blowers 1997).

Ecological modernization gives a key role to the relationship between government and industry, one that gives industry an important say in influencing the emerging agenda of sustainable development. The World Business Council for Sustainable

Development was formed for precisely this reason, and it is evident in the Type II partnership deals struck at the WSSD. However, there is a danger in this approach in that multinational corporations, globally responsible for extensive environmental degradation and resource depletion, can be cast as corporate environmentalists upon whom society can rely to promote sustainable development. At the same time, in trade relations, in the structures of international finance and in the generation of debt, the poor are cast as the perpetrators of environmental decline and as a barrier to a more progressive future. In addition, hidden from the view that the promotion of sustainable development is a mere technical task, associated with the application of indicators and the promotion of ecological modernization, is the fact that there are competing understandings of what sustainable development means and competing interpretations of what is needed to put that development model into practice.

Embedding the local in the global

Viewed from the local level, sustainable development is about promoting social change within the community, to take account of locally agreed upon ecological, cultural, political and social preferences. LA21 practices have helped to put flesh, as it were, on what this means in practice. Among the keys to identifying sustainable development priorities at the local level is the opening up of policy-making processes to wider groups within society and the economic sphere. However, this is not a simple task, as it requires a public that has learned a civic spirit and that no longer sees the public sphere as a forum for narrow self-interest.

The focus on the local level has also expanded the agenda of sustainable development. It has been extended into the way in which development exploits physical space, making land use in particular an increasingly important component of the sustainable development model. Promoting sustainable development requires integrating environmental considerations at the strategic level, especially into land-use planning. This, in turn, is premised upon willingness to dismantle interdepartmental rivalries in local authorities and to change existing institutional practices, and motivation to weaken entrenched policy coalitions (Bulkeley and Betsill 2005). New moves to create sustainable development at the urban level create the conditions not only for the integration of sustainable development considerations into the planning process, but also for enhancing the contribution that urban development makes to the construction of a sustainable future.

The model of sustainable development places a great deal of confidence in government at the local level as well as in the capacity of the sustainable development agenda to lead to social mobilization. The advantage of this emphasis is that it enhances the chances of generating examples of sustainable development practice.

As these successes become a tangible aspect of everyday life the model of sustainable development will acquire increased legitimacy and acceptance (Bridger and Luloff 1999). However, the discussions raised in this book, particularly as they relate to LA21, point to the necessity to pursue local development needs in ways that take account of the wider governance and ecological systems in which the local level is embedded.

Promoting sustainable development cannot rest on the weight or the input of traditional political authority alone, particularly that which is vested in national governments. The model of sustainable development promotes a governance process that engages state and non-state actors, the public and the private sectors, as they wrestle to agree priorities and devise action plans to put the commitment to sustainable development into practice through concrete development projects. Only through governance structures that are invigorated through the sense of partnership and shared responsibility, through the expression of empathy for the needs of the many over and above the wants of the few, and through the acceptance of humans as part of, not dominant over, nature, can the conditions be created to bring this development model to fruition.

Returning to the Brundtland formulation

The Brundtland formulation of sustainable development represents a radical agenda for social change. Whether it has been treated as such by the system of environmental governance that it has spawned is a separate issue. This book began by pointing out that the Brundtland definition now commands authoritative status, acting as a guiding principle of economic and social development. It ends by arguing that precisely because of the radical nature of its agenda, those that have engaged with the promotion of sustainable development have not adhered to all its principles or its recommended practices.

However, let us not throw the baby out with the bath water! Because international political and economic processes have restricted the agenda of sustainable development, it does not mean that the promotion of sustainable development is itself a limited agenda for change. Promoting a model of sustainable development recognizes that every human interaction with the world brings change, but it challenges society to find ways to ensure that these changes are for the betterment of all. This means that the promotion of sustainable development necessitates the adoption of a spirit of compassion not only for other human beings but also for all life forms. With such steps, a model of development can be constructed that opens up a future for the coming generations that will inherit the earth.

Bibliography

Achterberg, W. (1993) 'Can liberal democracy survive the environmental crisis? Sustainability, liberal neutrality and overlapping consensus', in A. Dobson and P. Lucardie (eds) *The Politics of Nature: Explorations in Green Political Theory*, London: Routledge, 81–101.

Agarwal, B. (1992) 'The gender and environment debate: lessons from India', *Feminist Studies*, 18(1): 119–58.

Agarwal, B. (1997) 'Gender perspectives on environmental action: issues of equity, agency and participation', in J.W. Scott, C. Kaplan and D. Keates (eds) *Transitions, Environments, Translations: Feminisms in International Policies*, London: Routledge, 189–225.

Anderson, T. (2002) 'The Cartagena Protocol on Biosafety to the Convention on Biological Diversity: trade liberalisation, the WTO, and the environment', *Asia Pacific Journal of Environmental Law*, 7(1): 1–37.

Arrow, K., Bolin, B., Costanza, R., Dasgupta, P., Folke, C., Holling, C.S., Jansson, B.-O., Levin, S., Mäler, K.-G., Perrings, C. and Pimentel, D. (1995) 'Economic growth, carrying capacity, and the environment', *Science*, 268: 520–1.

Baker, S. (1994) 'Structural adjustment and the environment: the gender dimension', in P. Rajput and H.L. Swarup (eds) *Women and Globalisation: Reflections, Options and Strategies*, New Delhi: Ashish Publishing House, 313–41.

Baker, S. (2000) 'The European Union: integration, competition, growth and sustainability', in W.M. Lafferty and J. Meadowcroft (eds) *Implementing Sustainable Development: Strategies and Initiatives in High Consumption Societies*, Oxford: Oxford University Press, 303–36.

Baker, S. (2002) 'Environmental politics and transition', in F.W. Carter and D. Turnock (eds) *Environmental Problems of East Central Europe*, 2nd edn, London: Routledge, 22–39.

Baker, S. (2003) 'The dynamics of European Union biodiversity policy: interactive functional and institutional logics', *Environmental Politics*, 12(3): 23–41.

Baker, S. (2004) 'The challenge of ecofeminism for European politics', in J. Barry, B. Baxter and R. Dunphy (eds) *Europe, Globalization and Sustainable Development*, London: Routledge, 15–30.

Baker, S. (2005a) 'Values and principles of EU environmental policy', in S. Lucarelli and I. Manners (eds) *Values in European Union Policy*, London: Routledge.

Baker, S. (2005b) 'The environmental dimensions of transition in Central and South East Europe', in I. Bell (ed.) *Central and South-Eastern Europe 2005*, London: Europa Publications, 53–63.

Baker, S. and McCormick, J. (2004) 'Sustainable development: comparative understandings and responses', in N.J. Vig and M.C. Faure (eds) *Green Giants? Environmental Policy of the United States and the European Union*, Cambridge, MA: MIT Press, 277–302.

Baker, S. and Welsh, I. (2000) 'Differentiating Western influences on transition societies in Eastern Europe: a preliminary exploration', *Journal of European Area Studies*, 8(1): 79–103.

Baker, S., Kousis, M., Richardson, D. and Young, S. (eds) (1997) *The Politics of Sustainable Development: Theory, Policy and Practice within the European Union*, London: Routledge.

Ball, T. (2000) 'The earth belongs to the living: Thomas Jefferson and the problem of intergenerational relations', *Environmental Politics*, 9(2): 61–77.

Banerjee, S.B. (2003) 'Who sustains whose development? Sustainable development and the reinvention of nature', *Organization Studies*, 24(1): 143–80.

Barnes, I. (1995) 'Environment, democracy and community', *Environmental Politics*, 4(4): 101–33.

Barry, J. (1999) *Environment and Social Theory*, London: Routledge.

Becker, E., Jahn, T. and Stieß, I. (1999) 'Exploring uncommon ground: sustainability and the social sciences', in E. Becker and T. Jahn (eds) *Sustainability and the Social Sciences: A Cross-disciplinary Approach to Integrating Environmental Considerations into Theoretical Reorientation*, London: Zed Books, 1–22.

Bell, D.R. (2004) 'Sustainability through democratization? The Aarhus Convention and the future of environmental decision making in Europe', in J. Barry, B. Baxter and R. Dunphy (eds) *Europe, Globalization and Sustainable Development*, London: Routledge, 94–112.

Bigg, T. and Dodds, F. (1997) 'The UN Commission on Sustainable Development', in F. Dodds (ed.) *The Way Forward: Beyond Agenda 21*, London: Earthscan, 15–36.

BirdLife International (2004) *Birds in the European Union: A Status Assessment*, Wageningeu, The Netherlands: BirdLife International.

Blowers, A. (1997) 'Environmental policy: ecological modernisation or the risk society', *Urban Studies*, 34(5/6): 845–71.

Bodansky, D. (2002) 'US climate policy after Kyoto: elements for success', paper presented at the BP Transatlantic Programme Workshop 'The Kyoto Protocol without America: Finding a Way forward after Marrakech', Florence: European University Institute, 21–22 June.

Braidotti, R., Charkiewicz, E., Häusler, S. and Wieringa, S. (1994) *Women, the Environment and Sustainable Development: Towards a Theoretical Synthesis*, London: Zed Books.

Breitmeier, H. (1997) 'International organizations and the creation of environmental regimes', in O.R. Young (ed.) *Global Governance: Drawing Insights from the Environmental Experience*, Cambridge, MA: MIT Press, 87–114.

Bretherton, C. and Vogler, J. (1999) *The European Union as a Global Actor*, London: Routledge.

Bridger, J.C. and Luloff, A.E. (1999) 'Toward an interactive approach to sustainable community development', *Journal of Rural Studies*, 15(4): 377–87.

Bryner, G. (2000) 'The United States: sorry – not our problem', in W.M. Lafferty and J. Meadowcroft (eds) *Implementing Sustainable Development: Strategies and Initiatives in High Consumption Societies*, Oxford: Oxford University Press, 273–302.

Buckingham-Hatfield, S. (2000) *Gender and Environment*, London: Routledge.

Bührs, T. (2003) 'From diffusion to defusion: the roots and effects of environmental innovation in New Zealand', *Environmental Politics*, 12(3): 83–101.

Bulkeley, H. and Betsill, M. (2005) 'Rethinking sustainable cities: multilevel governance and the "urban" politics of climate change', *Environmental Politics*, 14(1): 42–63.

Bureau of Oceans and International Environmental and Scientific Affairs, US Department of State (2002) 'Congo Basin Forest Partnership: U.S. Contribution', Fact Sheet, 12 August, http://www.state.gov/g/oes/rls/fs/2003/23208.htm.

Carter, F.W. and Turnock, D. (eds) (2001) *Environmental Problems of East Central Europe*, 2nd edn, London: Routledge.

CEC (Commission of the European Communities) (1973) 'First Community Action Programme on the Environment', *Official Journal*, C 112, 20 December.

CEC (Commission of the European Communities) (1977) 'Second Environmental Action Programme (1977–1981)', *Official Journal*, C 139, 13 June.

CEC (Commission of the European Communities) (1983) 'Third Environmental Action Programme', *Official Journal*, C 46, 17 February.

CEC (Commission of the European Communities) (1992) *Towards Sustainability: A European Community Programme of Policy and Action in Relation to the Environment (1992–2000)*, COM(1992)23–2 FINAL, Luxembourg: Office for Official Publications of the European Communities.

CEC (Commission of the European Communities) (1997a) *Communication from the Commission to the Council, the European Parliament, the Economic and Social Committee and the Committee of the Regions concerning the Energy Dimension of Climate Change*, COM(1997)196 FINAL, Luxembourg: Office for Official Publications of the European Communities.

CEC (Commission of the European Communities) (1997b) *Communication from the Commission to the Council, the European Parliament, the Economic and Social Committee and the Committee of the Regions: Climate Change – The EU Approach for Kyoto*, COM(1997)481 FINAL, Luxembourg: Office for Official Publications of the European Communities.

CEC (Commission of the European Communities) (1997c) *Agenda 21: The First Five Years: European Community Progress on the Implementation of Agenda 21 1992–1997*, Luxembourg: Office for Official Publications of the European Communities.

CEC (Commission of the European Communities) (1998a) *Sustainable Urban Development in the European Union: A Framework for Action*, COM(1998)605 FINAL, Luxembourg: Office for Official Publications of the European Communities.

CEC (Commission of the European Communities) (1998b) *Communication from the Commission to the Council and the European Parliament on a European Community Biodiversity Strategy*, COM(1998)42 FINAL, Luxembourg: Office for Official Publications of the European Communities.

CEC (Commission of the European Communities) (1998c) *Communication from the Commission: Strengthening Environmental Integration within Community Energy Policy*, COM(1998)571 FINAL, Luxembourg: Office for Official Publications of the European Communities.

CEC (Commission of the European Communities) (1998d) *Communication from the Commission to the Council and the European Parliament on Nuclear Sector Related Activities for the Applicant Countries of Central and Eastern Europe and the New Independent States*, COM(1998)134 FINAL, Luxembourg: Office for Official Publications of the European Communities.

CEC (Commission of the European Communities) (1999a) *Communication from the Commission to the Council, the European Parliament, the Economic and Social Committee and the Committee of the Regions: Directions towards Sustainable Agriculture*, COM (1999)22 FINAL, Luxembourg: Office for Official Publications of the European Communities.

CEC (Commission of the European Communities) (1999b) *Communication from the Commission to the Council and the European Parliament: Preparing for Implementation of the Kyoto Protocol*, COM(1999)230 FINAL, Luxembourg: Office for Official Publications of the European Communities.

CEC (Commission of the European Communities) (1999c) 'Commission working document – From Cardiff to Helsinki and beyond: report to the European Council on integrating environmental concerns and sustainable development into Community policies', SEC(1999)1941 FINAL, Brussels: European Commission.

CEC (Commission of the European Communities) (2000a) *Communication from the Commission on the Precautionary Principle*, COM(2000)1 FINAL, Luxembourg: Office for Official Publications of the European Communities.

CEC (Commission of the European Communities) (2000b) *Communication from the Commission – Europe's Environment: What Directions for the Future? The Global Assessment of the European Community Programme of Policy and Action in Relation to the Environment and Sustainable Development, 'Towards Sustainability'*, COM(1999)543 FINAL, Luxembourg: Office for Official Publications of the European Communities.

CEC (Commission of the European Communities) (2001a) *Communication from the Commission: The Challenge of Environmental Financing in the Candidate Countries*, COM(2001)304 FINAL, Luxembourg: Office for Official Publications of the European Communities.

CEC (Commission of the European Communities) (2001b) *Communication from the Commission to the Council, the European Parliament, the Economic and Social*

Committee and the Committee of the Regions on the Sixth Environment Action Programme of the European Community, 'Environment 2010: Our Future, Our Choice': The Sixth Environment Action Programme, COM(2001)31 FINAL, Luxembourg: Office for Official Publications of the European Communities.

CEC (Commission of the European Communities) (2001c) Communication from the Commission – A Sustainable Europe for a Better World: A European Union Strategy for Sustainable Development, COM(2001)264 FINAL, Luxembourg: Office for Official Publications of the European Communities.

CEC (Commission of the European Communities) (2001d) 'Commission staff working paper – Integrating environment and sustainable development into energy and transport policies: Review Report 2001 and implementation of the strategies', SEC(2001)502, Brussels: European Commission.

CEC (Commission of the European Communities) (2001e) Communication from the Commission to the Council and European Parliament – Ten Years after Rio: Preparing for the World Summit on Sustainable Development in 2002, COM(2001)53 FINAL, Luxembourg: Office for Official Publications of the European Communities.

CEC (Commission of the European Communities) (2002a) Communication from the Commission Setting out a Community Action Plan to Integrate Environmental Protection Requirements into the Common Fisheries Policy, COM(2002)186 FINAL, Luxembourg: Office for Official Publications of the European Communities.

CEC (Commission of the European Communities) (2002b) The Enlargement Process and the Three Pre-accession Instruments: Phare, ISPA, Sapard, Brussels: Commission of the European Communities.

CEC (Commission of the European Communities) (2002c) Road Maps for Bulgaria and Romania, COM(2002)624 FINAL, Luxembourg: Office for Official Publications of the European Communities.

CEC (Commission of the European Communities) (2002d) Industrial Policy in an Enlarged Europe, COM(2002)714 FINAL, Luxembourg: Office for Official Publications of the European Communities.

CEC (Commission of the European Communities) (2003) 'European economy', Occasional Paper 4, Brussels: Directorate General of Economic and Financial Affairs, Commission of the European Communities.

CEC (Commission of the European Communities) (2004a) Towards a Thematic Strategy on the Urban Environment, COM(2004)60 FINAL, Luxembourg: Office for Official Publications of the European Communities.

CEC (Commission of the European Communities) (2004b) 'Climate change: projections show EU on track to meet Kyoto Protocol emission targets', press release, IP/04/1522, 21 December.

CEE Bankwatch Network and Friends of the Earth Europe (2002) 'Billions of sustainability: lessons learned from the use of pre-accession funds', London: CEE Bankwatch Network and Friends of the Earth Europe.

Christoff, P. (1996) 'Ecological modernisation, ecological modernities', Environmental Politics, 5: 476–500.

Cohen, M.J. and Egelston, A. (2003) 'The Bush administration and climate change:

prospects for an effective policy response', *Journal of Environmental Policy and Planning*, 5(4): 315–31.

Couch, C. and Dennemann, A. (2000) 'Urban regeneration and sustainable development in Britain: the example of the Liverpool Ropewalks Partnership', *Cities*, 17(2): 137–47.

Curtin, D.M. (1999) 'Transparency and political participation in EU governance: a role for civil society?', *Cultural Values*, 3(4): 445–71.

Daly, H.E. (1977) *Steady-state Economics*, San Francisco, CA: Freeman.

Daly, H.E. and Cobb, J.B. (1990) *For the Common Good*, London: Green Print.

Department of the Environment, Transport and the Regions (1999) *A Better Quality of Life: A Strategy for Sustainable Development for the United Kingdom*, Cm 4345, London: HMSO.

Diamond, I. and Orenstein G.F. (eds) (1990) *Reweaving the World: The Emergence of Ecofeminism*, San Francisco, CA: Sierra Club Books.

Dobson, A. (1993) 'Critical theory and Green politics', in A. Dobson and P. Lucardie (eds) *The Politics of Nature*, London: Routledge, 190–209.

Dobson, A. (1996) 'Representative democracy and the environment', in W.M. Lafferty and J. Meadowcroft (eds) *Democracy and the Environment: Problems and Prospects*, Cheltenham: Edward Elgar, 124–39.

Dobson, A. (1998) *Justice and the Environment: Conceptions of Environmental Sustainability and Dimensions of Social Justice*, Oxford: Oxford University Press.

Dodds, F. (ed.) (1997) *The Way Forward: Beyond Agenda 21*, London: Earthscan.

Dresner, S. (2002) *The Principles of Sustainability*, London: Earthscan.

Dryzek, J. (1990) *Discursive Democracy*, Cambridge: Cambridge University Press.

Dryzek, J.S. (1992) 'Ecology and discursive democracy: beyond liberal capitalism and the administrative state', *Capitalism, Nature, Socialism*, 10: 18–42.

Eckerberg, K. (1999) 'Sweden: combining municipal and national efforts for quick progress', in W. Lafferty (ed.) *Implementing LA21 in Europe: New Initiative for Sustainable Communities*, Oslo: Prosus.

Eckerberg, K. and Forsberg, B. (1998) 'Implementing Agenda 21 in local government: the Swedish experience', *Local Environment*, 3(3): 333–47.

Eckerberg, K. and Lafferty, W. (1998) 'Comparative perspectives on evaluation and explanation', in W. Lafferty and K. Eckerberg (eds) *From the Earth Summit to Local Agenda 21: Working towards Sustainable Development*, London: Earthscan, 238–63.

Eckerberg, K., Baker, S., Marell, A., Dahlgren, M. and Wahlström, A. (2005) *LIP in Context: Central Government, Business and Comparative Perspectives*, Stockholm: Swedish Environmental Protection Agency.

The Ecologist (1993) *Whose Common Future? Reclaiming the Commons*, London: Earthscan.

EEA (European Environment Agency) (1995) *The Environment in the European Union 1995: Report for the Review of the Fifth Environmental Action EAP*, Copenhagen: EEA.

EEA (European Environment Agency) (1999) *Environment in the European Union at the Turn of the Century*, Copenhagen: EEA.

EEA (European Environment Agency) (2002a) 'Environmental signals: benchmarking the millennium', Environmental Assessment Report 9, Copenhagen: EEA.

EEA (European Environment Agency) (2002b) 'Paving the way for EU enlargement: indicators of transport and environment integration, TERM 2002', Environmental Issue Report 32, Copenhagen: EEA.

EEA (European Environment Agency) (2003) 'Europe's environment: the third assessment', Environmental Assessment Report 10, Copenhagen: EEA.

EEA (European Environment Agency) (2004) 'Lisbon goal poses choice between efficiency leap or exploiting the planet, EEA head warns', http://org.eea.eu.int/ documents/newsreleases/10thAnniversary-eu, accessed 1 December 2004.

Ehrlich, P. and Ehrlich, A. (1989) 'Too many rich folks', *Populi*, 16(3): 3–29.

Eichhammer, W. (2001) 'Integrating environment and sustainable development into energy policy: challenges for candidate countries', rapporteur's report of the Workshop DG TRENT/Taiex, Brussels, 24 October 2000, Fraunhofer Institute for Systems and Innovation Research, FhG-ISI.

Ekins, P. (2000) *Economic Growth and Environmental Sustainability: The Prospects for Green Growth*, London: Routledge.

Elliott, I. (2002) 'Global environmental governance', in R. Wilkinson and S. Hughes (eds) *Global Governance: Critical Perspectives*, London: Routledge, 57–74.

Escobar, A. (1995) *Encountering Development: The Making and Unmaking of the Third World, 1945–1992*, Princeton, NJ: Princeton University Press.

Esteva, G. (1992) 'Development', in W. Sachs (ed.) *The Development Dictionary*, London: Zed Books, 6–25.

European Community (2004) 'Treaty establishing a constitution for Europe', *Official Journal*, C 310, 16 December.

Fiorino, D.J. (1996) 'Environmental policy and the participation gap', in W.M. Lafferty and J. Meadowcroft (eds) *Democracy and the Environment: Problems and Prospects*, Cheltenham: Edward Elgar, 194–212.

Fisher, I. and Waliczky, Z. (2001) *An Assessment of the Potential Impact to the TINA Network on Important Bird Areas (IBAs) in the Accession Countries*, Sandy: BirdLife International/RSPB.

Foster, J.B. (1994) *The Vulnerable Planet: A Short Economic History of the Environment*, New York: Monthly Review Press.

Friends of the Earth Europe and CEE Bankwatch Network (2000) *Billions for Sustainability? The Use of EU Pre-accession Funds and their Environmental and Social Implications*, London: Friends of the Earth Europe.

Friends of the Earth UK (2002) 'US wrecks Earth Summit', press release, 4 September, http://www.foe.co.uk/resource/press_releases/20020904115910.html.

Gardiner, R. (2002) 'Governance for sustainable development: outcomes from Johannesburg', Stakeholder Forum for Our Common Future, Presentation to Global Governance 2002: Redefining Global Democracy, Montreal, Canada: Paper 8, Montreal: WHAT Governance Programme.

GATT (General Agreement on Tariffs and Trade) (1994) *Agreement Establishing the World Trade Organization*, Geneva: WTO.

Ghina, F. (2003) 'Sustainable development in small island developing states: the case of the Maldives', *Environment, Development and Sustainability*, 5: 139–65.

Gilbert, R., Stevenson, D., Giradet, H. and Stern, R. (1996) *Making Cities Work: The Role of Local Authorities in the Urban Environment*, London: Earthscan.

Goodland, R. and Daly, H. (1996) 'Environmental sustainability: universal and non-negotiable', *Ecological Applications*, 6(4): 1002–17.

Gowdy, J. (1999) 'Economic concepts of sustainability: relocating economic activity within society and environment', in E. Becker and T. Jahn (eds) *Sustainability and the Social Sciences: A Cross-disciplinary Approach to Integrating Environmental Considerations into Theoretical Reorientation*, London: Zed Books, 162–81.

Grin, J., Van de Graff, H. and Vergragt, P. (2003) 'Een derde generatie milieubeleid: een sociologisch perspectief en een beleidswetenschappelijk programma', *Beleidswetenschap*, 1: 51–72.

Grubb, M. and Gupta, J. (2000) 'Climate change, leadership and the EU', in J. Gupta and M. Grubb (eds) *Climate Change and European Leadership: A Sustainable Role for Europe*, Dordrecht: Kluwer.

Grubb, M., Koch, M., Munson, A., Sullivan, F. and Thomson, K. (1993) *The Earth Summit Agreements: A Guide and Assessment*, London: Earthscan.

Gupta, J. (2002) 'Global sustainable development governance: institutional challenges from a theoretical perspective', *International Environmental Agreements: Politics, Law and Economics*, 2: 361–88.

Haigh, N. (1996) 'Climate change policies and politics in the European Community', in T. O'Riordan and J. Jäger (eds) *Politics of Climate Change: A European Perspective*, London: Routledge, 155–85.

Hajer, M. (1995) *The Politics of Environmental Discourse: Ecological Modernization and the Policy Process*, Oxford: Oxford University Press.

Hall, S., Held, D. and McLennan, G. (1992) 'Introduction', in S. Hall, D. Held and T. McGrew (eds) *Modernity and its Futures*, Cambridge: Open University Press, 1–12.

Hass, P.M. (2002) 'UN conferences and constructivist governance of the environment', *Global Governance*, 8(1): 73–91.

Hayward, B.M. (1995) 'The greening of participatory democracy: a reconsideration of theory', *Environmental Politics*, 4(4): 215–36.

Hemmati, M. and Gardiner, R. (2001) 'Gender equality and sustainable development', Towards Earth Summit 2002, Social Briefing Paper 2, Washington, DC: Heinrich Böll Foundation.

Hens, L. and Nath, B. (2003) 'The Johannesburg Conference', *Environment, Development and Sustainability*, 5: 7–39.

Herkenrath, P. (2002) 'The implementation of the Convention on Biological Diversity: a non-government perspective ten years on', *RECIEL*, 11(1): 29–37.

Hill, C. (1993) 'The capability–expectations gap, or conceptualising Europe's international role', *Journal of Common Market Studies*, 31(3): 305–28.

Homeyer, I. von (2001) 'Enlarging EU environmental policy', paper presented to Environmental Studies Workshop, Florence: Robert Schuman Centre for Advanced Studies, European University Institute, May.

Huber, J. (1982) *The Lost Innocence of Ecology: New Technologies and Super-industrialized Development*, Frankfurt am Main: Fischer Verlag.

ICLEI (International Council for Local Environmental Initiatives) (2002) 'Local governments response to Agenda 21: summary report of Local Agenda 21 survey with regional focus', Toronto: ICLEI.

Iles, A. (2003) 'Rethinking differential obligations: equity under the Biodiversity Convention', *Leiden Journal of International Law*, 16: 217–51.

International Union for the Conservation of Nature and Natural Resources (1980) *World Conservation Strategy: Living Resources Conservation for Sustainable Development*, Gland: IUCN.

IPCC (1990) 'The IPCC 1990 first assessment overview and policymaker summaries', Geneva: IPCC.

Jacobs, M. (1991) *The Green Economy: Environment, Sustainable Development and the Politics of the Future*, London: Pluto Press.

Jacobs, M. (1995) 'Justice and sustainability', in J. Lovenduski and J. Stanyer (eds) *Contemporary Political Studies* III, Belfast: Political Studies Association of the UK, 1470–85.

Jagers, S.C. (2002) 'Justice, liberty and bread – for all? On the compatibility between sustainable development and liberal democracy', Gothenburg Studies in Politics 79, Gothenburg: Department of Political Science, Gothenburg University.

Jänicke, M. (1992) 'Conditions for environmental policy success: an international comparison', *Environmentalist*, 12(1): 47–58.

Jordan, A. (ed.) (2002) *Environmental Policy in the European Union: Actors, Institutions and Processes*, London: Earthscan.

Katz, E., Light, A. and Rothenberg, D. (2000) *Beneath the Surface: Critical Essays in the Philosophy of Deep Ecology*, Cambridge, MA: MIT Press.

Kenny, M. and Meadowcroft, J. (1999) 'Introduction', in M. Kenny and J. Meadowcroft (eds) *Planning Sustainability*, London: Routledge, 1–11.

Koch, M. and Grubb, M. (1993) 'Agenda 21', in M. Grubb, M. Koch, A. Munson, F. Sullivan and K. Thomson, *The Earth Summit Agreements: A Guide and Assessment*, London: Earthscan, 97–157.

Kooiman, J. (ed.) (1993) *Modern Governance: New Government–Society Interactions*, London: Sage.

Kraemer, R. Andreas (n.d.) *Results of the 'Cardiff Processes' – Assessing the State of Development and Charting the Way Ahead*, Berlin: Ecologic Institute for International and European Environmental Policy, Research Report no. 299 19 120 (UFOPLAN), abridged and translated from German.

Lafferty, W.M. (1995) 'The implementation of sustainable development in the European Union', in J. Lovenduski and J. Stanyer (eds) *Contemporary Political Studies* I, Belfast: Political Studies Association of the UK, 223–32.

Lafferty, W.M. (ed.) (2001) *Sustainable Communities in Europe*, London: Earthscan.

Lafferty, W.M. and Eckerberg, K. (1998a) 'The nature and purpose of Local Agenda 21', in W.M. Lafferty and K. Eckerberg (eds) *From the Earth Summit to Local Agenda 21: Working towards Sustainable Development*, London: Earthscan, 1–16.

Lafferty, W.M. and Eckerberg, K. (eds) (1998b) *From the Earth Summit to Local Agenda 21: Working Towards Sustainable Development*, London: Earthscan.

Lafferty, W.M. and Hovden, E. (2003) 'Environmental policy integration: towards an analytical framework', *Environmental Politics*, 12(3): 1–22.

Lafferty, W.M. and Langhelle, O. (eds) (1999) *Towards Sustainable Development: On the Goals of Development – and the Conditions of Sustainability*, London: Macmillan.

Lafferty, W.M. and Meadowcroft, J. (1996a) 'Democracy and the Environment: congruence and conflict – some preliminary reflections', in W.M. Lafferty and J. Meadowcroft (eds) *Democracy and the Environment: Problems and Prospects*, Cheltenham: Edward Elgar, 1–17.

Lafferty, W.M. and Meadowcroft, J. (1996b) 'Democracy and the environment: prospects for greater congruence', in W.M. Lafferty and J. Meadowcroft (eds) *Democracy and the Environment: Problems and Prospects*, Cheltenham: Edward Elgar, 256–72.

Lafferty, W.M. and Meadowcroft, J. (2000) 'Introduction', in W.M. Lafferty and J. Meadowcroft (eds) *Implementing Sustainable Development: Strategies and Initiatives in High Consumption Societies*, Oxford: Oxford University Press, 1–22.

Lake, R. (1998) 'Finance for the global environment: the effectiveness of the GEF as the financial mechanism to the Convention on Biological Diversity', *RECIEL*, 7(1): 68–75.

Langhelle, O. (2000) 'Why ecological modernisation and sustainable development should not be conflated', *Journal of Environmental Policy and Planning*, 2(4): 303–22.

Lefebvre, H. (1991) *The Production of Space*, Oxford: Blackwell.

Leggett, J.K. (2001) *The Carbon War: Global Warming and the End of the Oil Era*, London: Routledge.

Lélé, S. (1991) 'Sustainable development: a critical review', *World Development*, 19(6): 607–21.

Le Prestre, P.G. (2002) 'The CBD at ten: the long road to effectiveness', *Journal of International Wildlife Law and Policy*, 5: 269–85.

Maathai, W. (2003) *Green Belt Movement: Sharing the Approach and Experience*, New York: Lantern Books.

McManus, P. (1996) 'Contested terrains: politics, stories and discourses of sustainability', *Environmental Politics*, 5(1): 48–73.

Macnaghten, P. and Urry, J. (1998) *Contested Natures*, London: Sage Publications.

Mang, J. (ed.) (2000) *Root Cause of Biodiversity Loss*, London: Earthscan.

Martinez-Alier, J. (1999) 'The socio-ecological embeddedness of economic activity: the emergence of a transdisciplinary field', in E. Becker and T. Jahn (eds) *Sustainability and the Social Sciences: A Cross-disciplinary Approach to Integrating Environmental Considerations into Theoretical Reorientation*, London: Zed Books, 112–40.

Marvin, S. and Guy, S. (1997) 'Creating myths rather than sustainability: the transition fallacies of the new localism', *Local Environment*, 2: 311–18.

Meadowcroft, J. (1999) 'Planning for sustainable development: what can be learnt from the critics?', in M. Kenny and J. Meadowcroft (eds) *Planning Sustainability*, London: Routledge, 12–38.

Meadowcroft, J. (2002) 'The next step: a climate change briefing for European decision-makers', Policy Paper 02/13, Florence: Robert Schuman Centre for Advanced Studies, European University Institute.

Meadows, D.H., Meadows, D.L., Rander, J. and Behrens, W.W. (1972) *The Limits to Growth*, New York: Universe Books.

Merchant, C. (1980) *The Death of Nature: Women, Ecology and the Scientific Revolution*, New York: Harper & Row.

Merchant, C. (1992) *Radical Ecology: The Search for a Liveable World*, New York: Routledge.

Mol, A.P.J. (2000) 'The environmental movement in an era of ecological modernization', *Geoforum*, 31(1): 45–56.

Moltke, K. von (1997) 'Institutional interactions: the structure of regimes for trade and the environment', in O.R. Young (ed.) *Global Governance: Drawing Insights from the Environmental Experience*, Cambridge, MA: MIT Press, 247–71.

Moser, P. (2001) 'Glorification, disillusionment or the way into the future? The significance of Local Agenda 21 processes for the needs of local sustainability', *Local Environment*, 6(4): 453–67.

Murphy, J. (2000) 'Ecological modernisation', *Geoforum*, 31(1): 1–8.

Naess, A. (1989) *Ecology, Community and Lifestyle: Outline of an Ecosophy*, translated, edited and with an introduction by David Rothenberg, Cambridge: Cambridge University Press.

National Academy of Sciences (2001) *Science of Climate Change*, Washington, DC: National Academy Press.

OECD (Organization for Economic Cooperation and Development) (2003) 'Environmental indicators: development measurement and use', reference paper, Paris: Environment Directorate, Environment Performance and Information Division, OECD.

Oldfield, J.D. and Shaw, D.J.B. (2002) 'Revisiting sustainable development: Russian cultural and scientific traditions and the concept of sustainable development', *Area*, 34(4): 391–400.

Ordway, S.H. (1953) *Resources and the American Dream*, New York: Ronald Press.

O'Riordan, T. (1981) *Environmentalism*, 2nd edn, London: Pion Press.

O'Riordan, T. (1985) 'What does sustainability really mean? Theory and development of concepts of sustainability', *Sustainable Development in an Industrial Economy*, proceedings of a conference held at Queens' College, Cambridge, 23–25 June, Cambridge: UK Centre for Economic and Environmental Development.

O'Riordan, T. (ed.) (2001) *Globalism, Localism and Identity: New Perspectives on the Transition to Sustainability*, London: Earthscan.

Osborn, D. and Bigg, T. (1998) *Earth Summit II: Outcomes and Analysis*, London: Earthscan.

Osborn, F. (1953) *The Limits of the Earth*, Boston, MA: Little, Brown.

Ott, H. (2003) 'Warning signs from Delhi: troubled waters ahead for global climate policy', in G. Ulfstein and J. Werksman (eds) *Yearbook of International Environmental Law* XIII, *2002*, Oxford: Oxford University Press, 261–70.

Paehlke, R.C. (1989) *Environmentalism and the Future of Progressive Politics*, New Haven, CT: Yale University Press.

Paehlke, R.C. (1996) 'Environmental challenges to democratic practice', in W.M. Lafferty and J. Meadowcroft (eds) *Democracy and the Environment: Problems and Prospects*, Cheltenham: Edward Elgar, 18–38.

Paehlke, R. (2001) 'Environmental politics, sustainability and social science', *Environmental Politics*, 10(4): 1–22.

Pearce, D. (1994) *Blueprint 3: Measuring Sustainable Development*, London: Earthscan.

Pearce, D. (1995) *Blueprint 4: Capturing Global Environmental Value*, London: Earthscan.

Pearce, D. and Barbier, E.B. (2000) *Blueprint for a Sustainable Economy*, London: Earthscan.

Pearce, D., Markandya, A. and Barbier, E.B. (1989) *Blueprint for a Green Economy*, London: Earthscan.

Pepper, D. (1993) *Eco-Socialism: From Deep Ecology to Social Justice*, London: Routledge.

Pepper, D. (1996) *Modern Environmentalism: An Introduction*, London: Routledge.

Pepper, D. (1998) 'Sustainable development and ecological modernization: a radical homocentric perspective', *Sustainable Development*, 6: 1–7.

Pickvance, C. (2004) *Local Environmental Regulations in Post-Socialism: A Hungarian Case Study*, London: Ashgate.

Plumwood, V. (1986) 'Ecofeminism: an overview and discussion of positions and arguments', *Australasian Journal of Philosophy*, 64 (Supplement): 120–38.

President's Council on Sustainable Development (PCSD) (1996) *Sustainable America: A New Consensus*, Washington, DC: US Government Printing Office.

Prezzey, J. (1989) 'Definitions of sustainability', Working Paper 9, Cambridge: UK Centre for Economic and Environmental Development.

Rawls, J. (1971) *A Theory of Justice*, Cambridge, MA: Harvard University Press.

Reboratti, C.E. (1999) 'Territory, scale and sustainable development', in E. Becker and T. Jahn (eds) *Sustainability and the Social Sciences: A Cross-disciplinary Approach to Integrating Environmental Considerations into Theoretical Reorientation*, London: Zed Books, 207–22.

Redclift, M. (1987) *Sustainable Development: Exploring the Contradictions*, London: Routledge.

Redclift, M. (1997) 'Development and global environmental change', *Journal of International Development*, 9(3): 391–401.

Redclift, M. (1999) 'Dancing with wolves? Sustainability and the social sciences', in E. Becker and T. Jahn (eds) *Sustainability and the Social Sciences: A Cross-disciplinary Approach to Integrating Environmental Considerations into Theoretical Reorientation*, London: Zed Books, 267–73.

Redclift, M. and Woodgate, G. (1997) 'Sustainability and social construction', in M. Redclift and G. Woodgate (eds) *The International Handbook of Environmental Sociology*, Cheltenham: Edward Elgar, 55–70,

Reed, D. (ed.) (1996) *Structural Adjustment, the Environment, and Sustainable Development*, London: Earthscan.

Reed, D. (1997) 'The environmental legacy of Bretton Woods: the World Bank', in O.R. Young (ed.) *Global Governance: Drawing Insights from the Environmental Experience*, Cambridge, MA: MIT Press, 227–46.

Rees, W.E. (1999) 'The built environment and the ecosphere: a global perspective', *Building Research and Information*, 27(4/5): 206–20.

Regional Environmental Centre (2003) *Environmental Financing in Central and Eastern Europe, 1996–2001*, Budapest: Regional Environmental Centre.

Roseland, M. (2000) 'Sustainable community development: integrating environmental, economic, and social objectives', *Progress in Planning*, 54: 73–132.

Rostow, W. (1960) *The Stages of Economic Growth: A Non-Communist Manifesto*, Cambridge: Cambridge University Press.

Rowe, J. and Fudge, C. (2003) 'Linking national sustainable development strategy and local implementation: a case study of Sweden', *Local Environment*, 8(2): 125–40.

Ruffing, K. (2003) 'Johannesburg Summit: success or failure?', *OECD Observer*, 5 March.

Sachs, I. (1999) 'Social sustainability and whole development: exploring the dimensions of sustainable development', in E. Becker and T. Jahn (eds) *Sustainability and the Social Sciences: A Cross-disciplinary Approach to Integrating Environmental Considerations into Theoretical Reorientation*, London: Zed Books, 25–36.

Sachs, W. (1997) 'Sustainable development', in M. Redclift and G. Woodgate (eds) *The International Handbook of Environmental Sociology*, Cheltenham: Edward Elgar, 71–82.

Said, E. (1979) *Orientalism*, New York: Vintage Books.

Said, E. (1993) *Culture and Imperialism*, London: Vintage Books.

Satterthwaite, D. (1999) *The Earthscan Reader in Sustainable Cities*, London: Earthscan.

Secretariat CBD (Convention on Biological Diversity) (2002) *Report of the Sixth Meeting of the Conference of the Parties to the Convention on Biological Diversity*, UNEP/CBD/COP/6/20, New York: UNEP.

Shiva, V. (1989) *Staying Alive: Women, Ecology and Development*, London: Zed Books.

Shiva, V. (1991) *The Violence of the Green Revolution: Third World Agriculture, Ecology and Politics*, London: Zed Books.

Shiva, V. (1993) *Monocultures of the Mind: Perspectives on Biodiversity and Biotechnology*, London: Zed Books.

Shiva, V. (2000) *Stolen Harvest: The Hijacking of the Global Food Supply*, Cambridge: South End Press.

Simmonis, U. (1989), 'Ecological modernization of industrial society: three strategic elements', *International Social Science Journal*, 121: 347–61.

Simon, J.L. and Kahn, H. (1984) *The Resourceful Earth: A Response to Global 2000*, Oxford: Blackwell.

Smith, A. and Pickles, J. (1998) *Theorizing Transition: The Political Economy of Post-Communist Transformations*, London: Routledge.

Smith, M., Whitelegg, J. and Williams, N. (1998) *Greening the Built Environment*, London: Earthscan.

Spretnak, C. (1990) 'Ecofeminism: our roots and flowering', in I. Diamond and G.F. Orenstein (eds) *Reweaving the World: The Emergence of Ecofeminism*, San Francisco: Sierra Club Books.

Stark, D. (1997) 'Recombinant property in East European capitalism', in G. Grabher and D. Stark (eds) *Restructuring Networks in Post-Socialism*, Oxford: Oxford University Press.

Steblez, W.G. (2000) 'The mineral industries of Bulgaria and Rumania', in US Geological Survey, *Minerals Yearbook, 2000*, Reston, VA: USGS, available online, http://minerals.usgs.gov/minerals/pubs/country/2000/9408000.pdf, accessed 13 May 2005.

Sylvan, R. and Bennett, D.H. (1994) *The Greening of Ethics: From Anthropocentrism to Deep Green Theory*, Cambridge: White Horse Press.

Tickle, A. and Welsh, I. (1998) 'Environmental politics, civil society and post-communism', in A. Tickle and I. Welsh (eds) *Environment and Society in Eastern Europe*, Harlow: Longman, 156–85.

Tunney, J. (2004) 'The WTO and sustainability after Doha: a time for reassessment of the relationship between political science and law?', in J. Barry, B. Baxter and R. Dunphy (eds) *Europe, Globalization and Sustainable Development*, London: Routledge, 186–212.

UK Government (1999) *A Better Quality of Life – Strategy for Sustainable Development for the United Kingdom*, London: HMSO.

UN (United Nations) (2002a) *Outcome of the International Conference on Financing for Development, Monterrey Consensus, Monterrey, Mexico, 21 and 22 March 2002*, New York: United Nations.

UN (United Nations) (2002b) *World Summit on Sustainable Development: Johannesburg, 2002*, New York: UN Department of Economic and Social Affairs, Division for Sustainable Development.

UNDESA (United Nations Department of Economic and Social Affairs, Division for Sustainable Development) (2002) *The Johannesburg Summit Test: What Will Change?*, available at http://www.johannesburgsummit.org/html/whats_new/feature_story41.html, accessed 3 October 2003.

US Department of State (2002) 'United States climate action report', Washington, DC: US Government Printing Office, http://www.epa.gov/globalwarming/publications/car.

US Geological Survey (2000) *Mineral Yearbook*, Washington, DC: US Department of the Interior.

Vrolijk, C. (2002) 'International climate change management: the EU and other industrialised countries', paper presented at the BP Transatlantic Programme Workshop 'The Kyoto Protocol without America: Finding a way forward after Marrakech', Florence: European University Institute, 21–22 June.

Wackernagel, M. and Rees, W. (1996) *Our Ecological Footprint: Reducing Human Impact on the Earth*, Gabriola Island, BC: New Society Publishers.

WCED (World Commission on Environment and Development) (1987) *Our Common Future*, Oxford: Oxford University Press.

Weizäcker, E. von, Lovins, A.B. and Lovins, L.H. (1997) *Factor Four: Doubling Wealth – Halving Resource Use*, London: Earthscan.

Wilkinson, D. (2002) 'Jury out on sustainability impact assessment', *European Voice*, 8(30–1), October.

World Humanity Action Trust (2001) 'Governance programme, from governance for sustainable development', submission to the fourth meeting of the Open Ended Intergovernmental Group of Ministers or their Representatives on International Environmental Governance, Montreal, Canada, 30 November–1 December.

WTO (World Trade Organization) (2001) 'WTO Ministerial Conference: Fourth Session, Doha, 9–14 November 2001, Ministerial Declaration', WTO: WT/ min (01) /Dec /1, Geneva: WTO.

WWF (2000) 'The "new" European Union: WWF agenda for accession', Brussels: WWF Europe.

WWF (2002a) 'Enlargement and agriculture: enriching Europe, impoverishing our rural environment?', available online http://www.panda.org/downloads/europe/capen largementpositionpaper.pdf.

WWF (2002b) 'The "new" European Union', WWF agenda for accession: an update', available online http://www.panda.org/downloads/europe/agenda_summary.pdf.

WWF (2003) *Progress on Preparation for Natura 2000 in Future EU Member States*, Brussels: WWF Europe.

Yearley, S. (1997) 'Science and the environment', in M. Redclift and G. Woodgate (eds) *The International Handbook of Environmental Sociology*, Cheltenham: Edward Elgar, 227–36.

Young, O.R. (1997a) 'Rights, rules and resources in world affairs', in O.R. Young (ed.) *Global Governance: Drawing Insights from the Environmental Experience*, Cambridge, MA: MIT Press, 1–25.

Young, O.R. (ed.) (1997b) *Global Governance: Drawing Insights from the Environmental Experience*, Cambridge, MA: MIT Press.

Young, O.R. (2003) 'Environmental governance: the role of institutions in causing and confronting environmental problems', *International Environmental Agreements: Politics, Law and Economics*, 3: 377–93.

Index